Understanding Spectrum Liberalisation

Understanding Spectrum Liberalisation

Martin Sims

Toby Youell

Richard Womersley

CRC Press
Taylor & Francis Group
Boca Raton London New York

CRC Press is an imprint of the
Taylor & Francis Group, an **Informa** business

CRC Press
Taylor & Francis Group
6000 Broken Sound Parkway NW, Suite 300
Boca Raton, FL 33487-2742

First issued in paperback 2020

ISBN 13: 978-0-367-57557-1 (pbk)
ISBN 13: 978-1-4987-0502-8 (hbk)

Visit the Taylor & Francis Web site at
http://www.taylorandfrancis.com

and the CRC Press Web site at
http://www.crcpress.com

Contents

PART II　LIBERALISATION IN ACTION

PART IV   THE NEW AGENDA

Preface

What does this book try to do? We have two aims. Our first aim is to give an introduction to spectrum policy. Here we have in mind a civil servant or company employee who changes departments and needs to quickly grasp a puzzling new policy area.

Our second aim is to give an interpretation of recent developments in spectrum policy which will interest people already working in this field.

We were inspired by the recent history of spectrum policy, which is an intriguing story, and one that closely parallels the development of regulation in general. At around the same time that advanced economies pursued liberalisation at home and abroad, and Fukuyama hailed 'the end of history', regulators grew disenchanted with the 'civil servants-are-in-charge-of-the-airwaves' approach to radio spectrum management. Innovative regulators sought to liberalise radio spectrum – that is, to replace a technocratic and bureaucratic system with one that benefitted from the market's dynamism.

However, from around 2010, there was a creeping disappointment with the achievements of spectrum liberalisation and a new determination to explore spectrum sharing as a way of bringing dynamism to the sector.

Is this a revolution thwarted, with minds now turning to a new approach? Or are liberalisers regrouping after losing a few battles and may ultimately win the race? We hope that telling this story is an interesting way of introducing the subject to the uninitiated.

We also explain why spectrum liberalisation met with so many obstacles, and it is this analysis which we hope will interest current practitioners. History is about the concerns of the present, so the saying goes, and we hope our interpretation of the very recent past will stimulate debate about future policy.

Acknowledgements

Martin Sims

This has been a collaborative project so the first thanks are due to my co-authors, who have brought such an open-minded approach to all our discussions. Two of my *PolicyTracker* colleagues have been enormously valuable: Catherine Viola helped me to write Chapters 7 and 15 and Ross Bateson applied his long experience in satellite issues to Chapter 8. I am very grateful to four external experts who commented on individual chapters: Professor Gerard Pogorel, Philippa Marks of Plum Consulting, Professor Vincent Porter and Professor William Webb. My thanks also go to Tomoko Yamajo of KDDI in Japan whose support has allowed us to explore cutting edge policy issues in more detail.

Professor Webb was the co-author of *Essentials of Modern Spectrum Management*, which I found inspirational. I must also thank his collaborator on that book, Professor Martin Cave, who has been kind enough to share his insight and humour in *PolicyTracker* training courses for the past seven years, stimulating many of the ideas which have found their way into this book.

Finally, I owe a huge debt to my wife, Madhu Chauhan, for supporting me throughout this project. I value deeply her unique perspective on spectrum management.

Toby Youell

Spectrum is a fascinating, multidisciplinary and downright confusing subject to deal with. From engineering to economics, and from diplomacy to accountancy, professionals in the global spectrum management community manage to master quite different fields of human activity. In the last two years, these unsung masters of the universe have fielded my pestilent and sometimes naïve questions with wisdom and patience. I thank all the professionals who have taken the time to explain spectrum's many facets to me.

While researching for this book has led to some fruitful cross-fertilisation for everyday reporting in the spectrum world, it has also diverted an enormous amount of time away from my day-job reporting for the *PolicyTracker* newsletter. So I thank the editor of the newsletter, my boss and co-author, Martin, for tolerating and supporting this.

Most of all, thanks to my friends, family and my wife, who for some reason does not always find radio spectrum policy quite as fascinating as I do, despite my best efforts to convince them in conversation. Again, thank you for tolerating this.

Richard Womersley

First, thanks to my wife, Kate, who not only has to endure me 'playing radio' from time to time but who also has to keep our family afloat while I sail the seven seas, helping others around the world fathom out their radio spectrum problems.

My life-long journey through the radio spectrum began as a teenager in Yorkshire, and I shall ever be indebted to the members of the Sheffield Amateur Radio Club in the mid-1980s (Tom Haddon, Tony Whittaker, Norman Mason and many more) without whose encouragement, enthusiasm and patience I would not be where I am today.

Last, I owe an enormous amount of gratitude to my parents who had the insight to believe that allowing me to spend my spare time playing with computers and radios might lead to greater things. I hope I have done them proud.

Authors

Martin Sims is the founder and managing director of *PolicyTracker*, the spectrum management newsletter. He is also a consultant on spectrum policy issues, particularly relating to auctions.

Toby Youell is a journalist on *PolicyTracker* and has a degree in politics and Eastern European Studies from University College London, London, United Kingdom.

Richard Womersley is the director of Spectrum Consulting at LS telcom, based in Lichtenau, Germany, and has over 20 years of experience in spectrum management.

PART I
SETTING THE SCENE

1

INTRODUCTION

Since the mid 1990s, a wind of change has blown through the world of spectrum management, transforming the way engineers, administrators and politicians think about the use of the airwaves. There has been a concerted attempt to liberalise the sector (i.e. to make it more open to market forces), but these attempts have not been as successful as many had hoped. It has been suggested that the liberalisation agenda has achieved all it can, and a new paradigm is developing where the focus is sharing spectrum.

Perhaps we should back up for a moment. What is *spectrum*? Well, we use it to send radio signals to and from a huge range of devices. TV broadcasters use it to get a terrestrial signal to the TV aerial on your roof, radar uses the spectrum to get the positions of planes and soldiers use it on the battlefield. Satellite broadcasters use it to send a signal to your satellite dish, and mobile companies use it to get from your handset to the nearest base station. Those signals are going quite a long way, but we also use the spectrum for very short-range communication: wireless car keys, Bluetooth speakers and Wi-Fi in coffee shops are all examples of this.

Why do we care about spectrum? We care because it underpins current growth industries such as computing and the mobile sector and is vital to many visions of boosting future economic growth. The better our phones, tablets and laptops can connect to a central company database, the more efficient we can be. With an increasingly mobile workforce, this demands the use of spectrum rather than wired technologies. Future economic growth could come from technologies like machine-to-machine (M2M) communication that allows inanimate objects to communicate making businesses more efficient. Imagine being able to update the software in a car without having to recall each individual model to the garage. Like many other M2M applications, this needs to be done wirelessly if it is to be cost effective.

So if you want spectrum, how would you get it? This is what spectrum managers deal with and, historically, these managers were employed by governments.

The dominant approach to spectrum management before the arrival of liberalisation was known as *command and control*. If you wanted spectrum, you asked one of these government officials and they gave you some if no one else was using it. There was not usually a charge, but there were specific restrictions on power and technology to stop other users suffering interference.

Until the 1990s, almost all spectrum licences were allocated in this manner and generally at zero cost, even the first mobile licences that laid the foundation for multi-billion-dollar companies that dominate stock markets around the world.* Command and control seemed out of place in a world where spectrum usage was underpinning high-growth industries like mobile and computing, and liberalising the spectrum market seemed a more attractive option.

The key principle of spectrum liberalisation is that users should pay for spectrum in the same way that they would pay to use any other asset, such as land or fuel. There is then an incentive to use spectrum efficiently, so it should therefore flow to the users who can deliver what economists call the highest socio-economic value.

Land is often used as a metaphor for the radio spectrum, and we can compare the value of different types of land with different pieces of the spectrum. Land in the centre of a city is used for commerce rather than agriculture because the goods and services associated with city centres, such as retail, banking, insurance, accountancy and legal services, generate more profit per square mile than agricultural products.

The development of satellite towns outside major cites demonstrates the economic process behind change in the use of land. In the 1960s Sandton, near Johannesburg in South Africa, was known for its small holdings, but in the 1990s as the centre of Johannesburg went into decline, many companies moved their headquarters to Sandton,

---

* For example, Vodafone acquired the foundation of its business – spectrum in the UK 900 MHz band – in the 1980s at zero cost (http://licensing.ofcom.org.uk/radio-communication-licences/mobile-wireless-broadband/cellular-wireless-broadband/policy-and-background/history-of-cellular-services).

making it the country's premier financial district. This asset – land in Sandton – had moved to a higher value use because it could be bought, sold and repurposed.

Spectrum liberalisers would like frequencies to be bought and sold in the same manner, and their basic agenda is as follows:

- Sell or price spectrum
- Create licences that are technology and service neutral and allow for change of use
- Allow those licences to be traded in a secondary market
- Boost spectrum efficiency by sharing frequencies where possible

With these mechanisms in place, in theory the use of the airwaves should naturally flow to the users or uses that can extract the highest value.

In the 1980s and 1990s, the application of economic theory to the use of the airwaves was a new (or rediscovered*) concept, but it became an increasingly popular topic in universities and at regulators like the Federal Communications Commission (FCC) in the United States. By 2000, this was inspiring the worldwide spread of concrete liberalisation measures – the most popular being technology neutral licencing and spectrum auctions. The United States led the field, with the United Kingdom following and the European Commission making liberalisation the cornerstone of its developing spectrum policy from the early years of the new millennium. Within a few years, all EU countries were required to allow trading and practice technology neutral licencing in the most valuable bands.

Putting Theory into Practice

Spectrum liberalisation has worked best when it has been applied to the sale of mobile spectrum. The world took notice when mobile licences were sold for hundreds of millions of dollars in the United States in the mid-1990s. The valuable nature of the asset was plain to see and confirmed when European countries sold 3G licences for

---

* See discussion of Coase, R. H. (1959). The Federal Communications Commission. *Journal of Law and Economics*, Vol. 2, in Chapter 2.

the equivalent of billions of dollars in 2000. That purest of market mechanisms – the auction – has proved to be a much more successful way of assigning spectrum than random processes like lotteries or beauty contests where the regulator chooses the best applicant.

However, around 2009 there was a major setback for the liberalisation project. Demand for mobile data had grown exponentially over the preceding years, leading the mobile industry and regulators to conclude that there was a pressing need for more spectrum.

The collective response to meeting this need did more than anything else to publicly signal the limits of spectrum liberalisation. Mobile companies did not directly seek to buy spectrum from other users, like broadcasters and the public sector, because in most countries this would not be practically possible. This was true even in the most liberalised markets, such as that of the United States.

Regulators did not tell mobile operators to source this spectrum through the market. They tried to reallocate the airwaves at a national level as well as sought international solutions by forming common policies at the regional and International Telecommunication Union levels.

It was a tacit admission either that liberalisation alone could not deliver, or that it was inherently a slow process, unfit for its biggest challenge so far.

A less dramatic failure for liberalisation has been the limited impact of spectrum trades in Europe. Trades have been allowed in most European countries for several years, but there are few examples of a dynamic secondary market, and practically no trades in the highest value mobile bands. This is puzzling, as there have been more trades in the United States and Australia.

Increasing interest in wholesale networks is another problematic issue for the liberalisation agenda. Economists had welcomed the development of mobile networks – separate infrastructures that are inherently more competitive than the single networks in fixed-line telephony. However, mobile operators have increasingly shared infrastructure to reduce costs, and the rise of the most recent mobile technology, Long Term Evolution (LTE), is accentuating this drive. Many argue that single, shared LTE networks are more efficient than multiple networks because of the large contiguous blocks of spectrum necessary to make them as effective as possible.

In countries such as Mexico, state-backed wholesale networks are seen as a solution to excessive concentration in the mobile industry. Several African countries are also proposing a wholesale network. This upsets the traditional regulatory model for mobile and pushes it more towards the fixed-line model where the key issue is regulated access to a single network.

There have also been concerns about the effectiveness of auctions as assignment mechanisms. Do they suck investment funds out of the industry when the greater economic need is to ensure the rapid roll-out of 4G networks? Do they perpetuate oligopoly, failing to recognise that mobile operators can give a greater value to excluding a competitor than actually using the spectrum to provide services?

As confidence has diminished in the ability of liberalisation alone to deliver an economically efficient spectrum market, policymakers' focus has shifted to other mechanisms, principally sharing and cognitive radio, in what has been dubbed a 'third phase' of spectrum management.

The interest in sharing has been spurred by the development of the Licensed Shared Access (LSA) concept, a legally standardised sharing agreement promoted by technology giants NSN and Qualcomm and endorsed by the US communications regulator, the FCC and the European Commission. The sharing can be specified by time, frequency and location and enforced by software that is already under development. This approach has been approved for use in the European Union and in the United States.

The Structure of This Book

This book is divided into four sections. In the first, we introduce the topic of spectrum liberalisation by explaining in more detail what it is and how it came about (Chapter 2); we also explain briefly how spectrum works in a technical sense (Chapter 3).

In Section II, 'Liberalisation in Action', we consider how spectrum liberalisation has been put into practice, looking first at spectrum auctions in Chapter 4, then moving on to liberalised licencing, public-sector spectrum, broadcasting, satellite, ultra wide band, Wi-Fi and whitespace in the succeeding chapters. In each chapter, we try to explain why liberalisation was an attractive option, how the policies were implemented and to what extent they were successful.

In Section III, 'The Limits of Liberalisation', we analyse why liberalisation has not been as successful as many hoped. This section starts with an analysis of liberalisation's failure to find more spectrum for mobile (Chapters 13 and 14), and then considers the patchy take-up of spectrum trading (Chapter 15). Chapter 16 investigates the implications of moves away from infrastructure competition and towards wholesale mobile networks. Chapter 17 examines the influence of the wider political context in which spectrum liberalisation operates. In Chapter 18, we discuss the criticisms of auctions. In analysing the successes and failures of liberalisation, we hope to highlight the characteristics and structures of the spectrum market.

Section IV, 'The New Agenda', considers in more detail the new enthusiasm for spectrum sharing, looking at LSA in Chapter 20 and cognitive radio in Chapter 21. In Chapter 22, we examine the likely implications of other new technologies for spectrum management.

The introductory and concluding chapters of Section IV (Chapters 19 and 23) consider whether we are really moving into a new phase of spectrum management. This debate has an ironic aspect: the 'new phase' focuses on sharing, but sharing was always a part of the liberalisation agenda. Is this best described as a new paradigm for spectrum management, or is it just a change of emphasis? We assess whether the key ideas of liberalisation – auctions, pricing, trading and liberalised licencing – are really obsolete or instruments in the regulatory toolbox that can only achieve so much and must be complemented by other measures.

2
THE PROMISE OF
LIBERALISATION

In this book, we argue that spectrum management has entered a new phase. Confidence in liberalisation as a cure-all is eroding, its limitations are becoming apparent and it no longer seems able to solve all the most pressing problems. Interest is growing in other perspectives.

But in the same way, that new approaches to spectrum management are a reaction to the weaknesses of liberalisation, spectrum liberalisation was itself a reaction to the previous orthodoxy, usually called *command and control*. Knowing more about the administrative approach to spectrum assignment, as it is more properly known, will help us understand why liberalisation was such an appealing prospect.

A History of Spectrum Policy

The command-and-control approach dates back to the early days of wireless communication at the start of the twentieth century. In this period, the new medium was starting to be used for communication with ships and as a means of military communication, and the main concern was to prevent interference.

But the spectrum policy approaches which we are still debating today, were already surfacing over a hundred years ago. Since the inception of wireless communications, there has always been pressure from agencies of the state to control the use of the airwaves. In the United Kingdom in the run-up to World War I, there was widespread concern about national security and the government did not want radio transmitters to be used by spies. So the Wireless Telegraphy Act was passed in 1904 requiring anyone with a transmitter or receiver to have a licence.

In the United States, the Navy made several attempts to control the airwaves because amateur users were interfering with communications to ships. However, bill proposals in 1910, 1917 and 1918 were rejected because lawmakers were concerned about giving so much power to the state and the departments of the state.*

This was the command-and-control approach in its infancy, and what is now known as the commons approach was there as well. The early radio amateurs were sending signals through the air. As governments did not restrict the amount of air we can breathe, the amateurs saw no reason for them to control this other use of a free natural resource.

These early radio amateurs saw spectrum as a 'commons': a free public space which could be used by anyone without charge or restriction. This approach to spectrum management continues today, most famously in 2.4 GHz, the main Wi-Fi band where permitted powers are so low that the danger of interference is very small.

But amateurs were not content with unpaid experimentation, and in the United States, as the 1920s progressed, many went from merely sending messages to playing music, talk-based entertainment and even forms of advertising. Despite operating out of garden sheds, they were becoming what we now know as radio stations.

To be commercially viable, these new stations needed certainty about their access to the airwaves. In short, they needed a licence, which would guarantee access for a defined period without interference. In the United Kingdom in 1920, Marconi, that pioneer of commercialising new wireless technologies, in association with one of the country's biggest newspapers, sought a licence from the military to transmit music. It was granted, and then withdrawn, and then awarded again when the generals were overruled by the government.

Here we see two aspects of current spectrum management developing. Commercialisation of the airwaves requires licenses, and those licences are being administered by a command-and-control regime.

The father of spectrum liberalisation, Ronald Coase, makes some keen observations about these early twentieth-century policy debates.

People were confused, he said, because they did not know quite how to categorise spectrum, and that led to confused policy making.

Spectrum was not a physical thing – like air we cannot hold, see, smell or touch it. Was it therefore a free public resource, like air? Or was

---

* See Coase, R. H. (1959). The Federal Communications Commission. *Journal of Law and Economics*, Vol. 2, p. 2.

it a means of delivering information and entertainment? It clearly was, but it had no physical existence like the paper of a newspaper or magazine. Was it a means of expression, and therefore entitled to the free speech protection of the First Amendment? In a sense yes, but this freedom needed restrictions: while a city might support 100 different newspapers and magazines, the airwaves did not necessarily have the space for this many radio stations. Would broadcasting have immense cultural significance, therefore meriting public support like libraries and museums? This may be possible, although this cultural aspect was not fully developed in the early 1920s, leading to a 'wait-and-see' attitude among policymakers.

'It was in the shadows cast by a mysterious technology that our views on broadcasting policy were formed', wrote Ronald Coase in 1959 in the seminal paper,[*] which is the foundation for the modern spectrum liberalisation movement. '[I]nterpreting the law on this subject is something like trying to interpret the law of the occult. It seems like dealing with something supernatural. I want to put it off as long as possible in the hope that it becomes more understandable', said William Howard Taft, Chief Justice of the US Supreme Court from 1921 to 1930 – the formative period for the broadcasting industry.[†]

For Coase, however, the answer was simple. Spectrum was a commercial input like any other and should be treated in the same way. For the past 30 years, the United States had got it wrong, he argued, and this all dated back to a legal judgement in 1926.

1926: The Era of Broadcasting Chaos

In the United States, the early 1920s were like the dotcom boom of the late 1990s. The radio industry really took off in 1922, going from 60 stations at the beginning of the year to over 560 at the end.[‡] Companies that wanted a slice of this boom industry were not afraid to challenge the government in the courts.

[*] Coase, R. H. (1959). The Federal Communications Commission. *Journal of Law and Economics*, Vol. 2.

[†] Ibid., p. 28.

[‡] Ibid., p. 4.

Under a law from 1912, people using radio equipment were required to have a licence, broadcast within a limited frequency range and give details of the wavelengths they used, but this gave the US government no specific power to prevent interference. Subsequent legal rulings did little to give the government clearer legal powers and matters came to a head in April 1926. The Zenith Radio company was taken to court for broadcasting outside the limits of what it felt was a restrictive licence. The case was thrown out and the attorney general subsequently confirmed the ruling that the government had no power to restrict radio stations' use of frequencies to prevent interference.

This meant that the government had to issue licenses to anyone who applied, and the broadcasters could decide when and where they broadcast; their power levels and hours of operation and their wavelengths, as long as they were within the 600- to 1600-m limits.

The 'Era of Broadcasting Chaos' had begun. Anyone could get a licence and more than 200 stations were set up in the next 9 months. Stations turned up the power to blast out their neighbours and started broadcasting on whatever frequencies they chose. It was chaos: commercial stations interfered with each other and with emergency communications.

In the previous 10 years, many attempts to tackle this issue had been abandoned but now the US government was forced to act. By December 1926, Congress had passed a law which set up the forerunner of the Federal Communications Commission (FCC) and the government had new powers to

- Assign wavelengths
- Determine power and location of transmitters
- Determine type of apparatus used
- Determine nature of service
- Make regulations to prevent interference
- Rule on suitability of licence holders

It was the start of the command-and-control regime as a way to regulate the commercial use of the airwaves, and crucially it was quite specific in one area: the airwaves belonged to the state, broadcasters did not own them, they merely had a licence to use them and they were not allowed to trade these licences.

What Went Wrong?

For Coase, policymakers looked at the Era of Broadcasting Chaos and drew exactly the wrong conclusion: commercial companies could not be trusted to use the airwaves responsibly without regulatory supervision.

The real problem, he argued, was the lack of property rights. The Zenith Case was like giving farmers access to land without specifying what the boundaries were. Cattle were being allowed to roam over fields which other farmers thought they owned because there were no boundary lines on a map to show where the fences could be built. Clearly defined property rights backed by the rule of law are the foundation on which modern economies are built, said Coase, and if broadcasters had been given those rights, there would have been no chaos.

Broadcasters just needed the right to broadcast in specific frequencies, over a specified coverage area, and at particular times. If they suffered interference, this could be resolved either by negotiation or ultimately by legal action.

Spectrum was not 'mysterious' and did not need to be treated in a special way where a newly created body had the power to withhold or award licences depending on the 'suitability' of the applicants. Broadcasters should own their spectrum and be able to trade it, said Coase, with similar property rights to land. In fact, the market already recognised the value of spectrum, he argued, because TV stations were sold at far more than the cost of their facilities and technical equipment. The considerable extra value came from their licence to access the spectrum.

Watching from across the Water

The Era of Broadcasting Chaos was influential not just in the United States but in the United Kingdom as well. A post office official, F.J. Brown, visited the United States and wrote a report which was instrumental in setting up the British Broadcasting Corporation (BBC). 'The British government realised on the basis of the US experience that broadcasting was a new kind of resource whose management demanded a new kind of administration', according to a leading historian,* which is exactly the wrong conclusion according to Coase.

---

* Curran, J and Seaton, J (1997). *Power Without Responsibility*, p. 115 (Routledge).

The US experience and brief experiments like the Marconi daily broadcast in 1920 convinced the UK committee charged with setting up a broadcasting network that this was a natural monopoly. Monopolies would be abused by the commercial sector, and there were concerns about direct state control of such a powerful medium. So, the BBC, an independent body, operating under a royal charter, similar to the Bank of England's, was hit upon as the ideal way to operate and regulate broadcasting spectrum.

Similar solutions involving some form of arms-length or state intervention in broadcasting were adopted in many European countries. The European political tradition is less mistrustful of the state than the US political tradition, and this was reflected in their spectrum policy.

Coase's Legacy

The breadth and prescience of Coase's vision explain the continuing importance of this paper. His manifesto was not confined to the commercial use of the spectrum: he thought liberalisation should apply to the public sector as well. He recommended opportunity cost pricing for spectrum used by government bodies, something that was not adopted for almost 40 years:

> If the use of a frequency which if used industrially would contribute goods worth $1 million could be avoided by the construction of a wire system or the purchase of reserve vehicles costing $100,000, it is better that the frequency should not be used, however essential the project. It is the merit of the pricing system that, in these circumstances, a government department (unless very badly managed) would not use the frequency if made to pay for it.*

It is interesting to note that Coase's 1959 paper was not particularly influential at the time. Having conducted a high-level analysis of spectrum issues, he moved onto other issues, but when he won the Nobel Prize in 1991 for his work on companies, there was new interest in his earlier work. This coincided with the growth of mobile demonstrating the commercial value of the airwaves: his vision was a road map for this new future.

---

* Coase, R. H. (1959). The Federal Communications Commission. *Journal of Law and Economics*, Vol. 2, p. 21.

What Is Command and Control?

What do we actually mean by a command-and-control approach to spectrum management?

From a practical perspective, it is all about licensing. If I wanted to operate a new wireless service I would ask officials from the government or regulator and they would tell me whether I could have a licence. This licence would specify what technologies I could use, in what frequencies, at what power and, in some cases, at what times I could use them. These officials would have worked out whether my use of this frequency would cause interference to other uses and adjusted power limits and perhaps timing accordingly.

A good example of the command-and-control approach is the 1987 GSM Directive, which was an unusual piece of spectrum legislation because it was passed at the European, rather than at the national level. It required that all European Union (EU) countries should use Global System for Mobile (GSM) communication technology in the most popular mobile bands, namely, 900 and 1800 MHz. The idea was to overcome the problem of phones not working in other countries, but once GSM had been superseded as a standard this delayed mobile operators who wished to deploy superior technologies like Long Term Evolution or (4G) in their key spectrum assets.

To continue the comparison with land, under the command-and-control approach you are given spectrum by the government or regulator if there is some available. You do not buy it from a private owner, as you would with land. You are told what you can use the spectrum for and what technology you can deploy, rather like a farmer being told that he or she must grow potatoes and use a certain type of plough on newly acquired land.

The benefit of the command-and-control approach is that it prevents interference, but the disadvantage is its lack of flexibility. What if the farmer could make more money growing grain, or the mobile operator wants to use a superior technology?

Regulators Picking Winners

The command-and-control approach to assigning spectrum is inherently limiting. It is regulators or governments who make the decision about the best technology, not markets, and this is an approach which has been widely discredited by the 1990s.

A key nail in the coffin was the Multiplexed Analog Components (MAC) satellite TV standard mandated by the EU in 1986. Designed to end the multiplicity of standards in the 1950s and 1960s, this was put together by a European committee of experts with much public funding. MAC offered sharper pictures, digital stereo sound and widescreen, but the existing Phase Alternating Line (PAL) system was cheaper and easier. It also had the backing of the Murdoch empire as the chosen delivery platform for their Sky satellite service launched in 1988.

In short, MAC had superior features but the market preferred value over quality. Large amounts of public money had been wasted, and the conclusion drawn was that regulators should not be picking winners but where possible should let the market decide.

Another example was the European Radio Messaging System (ERMES) advanced paging system, for which the EU reserved spectrum in 1990. This offered advanced features like longer messages and voicemail, but it was no match for the emerging mobile industry which by 1993 was starting to offer SMS as well as voice. The spectrum reserved for ERMES remained largely unused until 2005 when the European Directive was repealed.

Reversing national legislation takes a long time, but undoing European legislation takes even longer: had the EU let the market decide about the merits of advanced paging systems, the 169 MHz would not have been largely unused for 15 years.

By the 1990s, leaving technology choice up to the market was considered best practice across information and communication technology policy, so there was considerable pressure to adopt this in spectrum assignment. However, whether regulators should mandate standards is not a simple black and white issue, as the success of the GSM Directive shows (see Box 2.1).

In fact, there is a partial solution to this problem. From the new millennium onwards, many regulators started awarding licences on a technology neutral basis. At its simplest, this means removing the word 'GSM' (or whichever technology) from the licences but using technical stipulations which mirror those of GSM. This allows operators to update their technologies, but only if the updates use the spectrum in

BOX 2.1 THE GSM DIRECTIVE: THE OTHER SIDE OF THE STANDARDS ARGUMENT

In limited circumstances, mandating standards does seem to be successful, and the GSM Directive was an example of this. The European mobile sector enjoyed a lead over the United States that lasted for 20 years: adoption rates were higher and prices were lower, and this was partly due to the single equipment market created by the 1987 GSM Directive. In contrast competing mobile standards in the United States meant that national roaming was not always possible. There were other factors at work, such as the adoption of the unpopular receiving party pays system in the United States, but the mandating of the GSM Directive is widely acknowledged as a cornerstone of the Europe mobile industry (Figure 2.1).

So why did mandating a technology work in mobile but not in TV or paging? The use of analogue cellular networks from the mid-1980s demonstrated a huge latent demand, with waiting lists years long to get a phone. This was a small market with development potential, rather than an established market like paging or satellite TV. GSM was the first digital cellular technology, the development which allowed mobile phones to become a mass

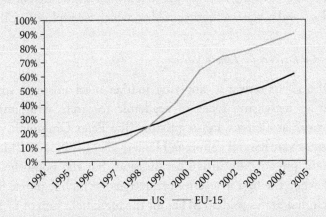

Figure 2.1 Mobile penetration 1994 to 2004. (From GSMA, *Optimising spectrum for future mobile service* needs based on data from FCC, Merrill Lynch, GSMA, London, UK, 2006. http://www.europa.eu.int. With permission.)

market product. Its main competition, the first Code Division Multiple Access (CDMA) standard developed by Qualcomm in the United States was not conceived of until 1988 – a year after the GSM Directive – and was not standardised until 1993.*

The lesson is that regulatory mandating of standards can work when it relates to a new market – such as the expansion of mobile from a niche offering to become a mass consumer product. If this new market takes off, the intervention can be very successful. But where governments have tried to influence an existing market – such as satellite TV or paging – it is a risky proposition. Established industries can usually bring new products to market themselves where they will be accepted or rejected based on a large number of factors ranging from price, quality and marketing, to the ability to bundle with other products.

The success factors are often too complex to predict, and by intervening the regulator risks wasting public money and foisting an unwanted product on the consumer, with consequent economic damage.

---

* CDG. CDMA History. https://www.cdg.org/resources/cdma_history.asp

the same way. Creating licences which allow for the desired flexibility in technology and usage is a thorny problem which we explore in Chapter 5.

Commercial Drivers for Liberalisation

From the 1990s onwards, applying market mechanism to spectrum became an increasing focus of academic research, with important papers from academics and regulators like Peter Crampton, Arthur De Vany, Evan Kwerel, Thomas Hazlett and Eli Noam. This was a significant pressure towards liberalisation, but the bombshell came in 1995 when the first in a series of auctions for mobile phone licences in the United States sold 30 MHz for the incredible sum of $7 billion. Two further auctions in the next two years raised $10 billion and $2.5 billion, respectively.

At the time, these were enormous figures, unheard of prices to pay for access to an invisible entity which had previously been the domain of an obscure band of engineers and hobbyists. Governments around the world sat up and took notice: Had they been sitting on an asset that was worth billions? Here was proof that a market-based approach to spectrum could pay huge dividends for the treasury and stimulate a rapidly growing new industry.

Mobile was that once-in-a-generation rapid growth industry which treasuries dream of – for the boost it gives the economy, the jobs it creates and the tax revenues it generates. What plastics had been in the 1960s and 1970s, and what computers had been in the 1970s and 1980s, mobile would become in the mid-1990s and beyond.

Figure 2.2 considers the example of Vodafone, which started with a UK analogue cellular licence in 1982, creating the country's first GSM mobile phone network in 1991 and eventually expanding into over 60 countries.

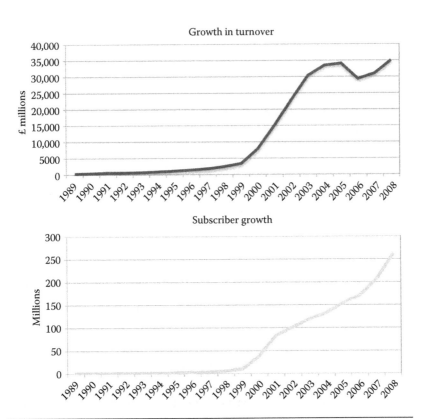

Figure 2.2 The growth of Vodafone from 1989 to 2008. (From Vodafone annual reports, Newbury, Berkshire, United Kingdom. With permission.)

The commercial value of spectrum was further underlined in 2000 with the infamous 3G auctions in Europe. The UK 3G auction in April 2000 raised the equivalent of $35 billion or 2.5 per cent of the country's GDP, the Germany auction in August raised $50 billion.

The Liberalisation Agenda

So what would a liberalised spectrum policy consist of? The first principle is to sell spectrum wherever possible. This has seen auctions becoming the default assignment mechanism in nearly every country for the higher value licences, principally mobile. This is the most widely adopted feature of liberalisation, and it is examined in more detail in Chapter 4. A smaller number of countries have also priced spectrum being used by the public sector, rather than giving it away for free, and we discuss this in Chapter 6.

Secondary trading, or allowing companies to buy and sell the right to use the airwaves with minimal regulatory intervention, is popular in the United States and Australia, allowed in the EU but not widely used elsewhere.

Trading has been a key part of the liberalisation agenda as it involves seeing spectrum as a 'normal' business input. If you can buy spectrum directly from the person currently using it, rather than having to go to the state or a government agency, then it encourages the wireless community to see spectrum as something with a specific and exchangeable value. This approach seems to work well in some countries and sectors and not others, and we examine this paradox in Chapter 15.

In this book, we argue that confidence is declining in liberalisation as a cure-all for the ills of the spectrum world and attention is moving to new approaches based on sharing. While this is true, sharing was always a part of the liberalisation agenda.

The huge success of Wi-Fi in the early 2000s boosted the case for having a variety of spectrum access regimes. While the mobile bands were clearly hugely valuable, an enormous industry had been built on the back on unlicensed spectrum – in the case of Wi-Fi principally the 2.4 GHz band. Having parts of the spectrum which adopt a 'commons' model means that barriers to entry are greatly lowered, stimulating innovation and new services. In the mobile bands, you would need to pay millions just to get access to the spectrum. In the unlicensed bands spectrum access is free, although the development costs remain. The rise of Wi-Fi is covered in more detail in Chapter 10.

It should also be noted that the liberalisation agenda only called for licensing *where necessary*, usually to prevent interference. Wi-Fi works on low power and so experiences little interference. In the higher frequency bands, for example 60 GHz, signals will travel only a few metres, meaning this band is ideal for applications like the wireless connection of a satellite box to a TV receiver. To require licences where the natural properties of the spectrum prevent interference would be perverse and economically damaging.

Liberalisers see sharing as a way of getting more economic value out of the available spectrum and had been involved in promoting technologies based on this concept, such as ultra wide band and whitespace access. These are discussed in Chapters 9 and 11.

However, in the future sharing is expected to take on a much greater importance, and we explain what that means in practice in the final section of the book.

3

A SPECTRUM ENGINEERING PRIMER

Introduction

What is the radio spectrum? At a basic level, it is an electromagnetic wave, consisting of both an electrical element and a magnetic element. It is part of the overall electromagnetic spectrum together with infrared, visible light, x-rays and gamma rays, as shown in Figure 3.1. It is a wave that travels through space at the speed of light. But the most important question in the context of spectrum liberalisation is the following: What gives the radio spectrum economic value? For this, we need to understand some of the basic technical characteristics of radio waves as it is these characteristics which give the radio spectrum value and which cause certain parts of the spectrum to have a higher value than others for particular applications.

When we talk about the radio spectrum, we generally use different frequencies to describe different parts of the spectrum. Frequencies are measured in Hertz (Hz), after the German physicist Heinrich Hertz who conducted the first experiments that proved that radio waves existed. But the frequency of a radio wave is intimately connected to its wavelength, by a very simple formula. If you multiply the frequency (in Hertz) by the wavelength (in metres), the result is always the speed of the light. Thus, we could equally talk about the wavelength of a radio signal instead of its frequency as the two are directly related. If we think of a radio wave in terms of its wavelength instead of its frequency, many of the characteristics we need to understand become more straightforward to conceptualise.

Different parts of the radio spectrum are banded together and given different names. Table 3.1 shows the recognised international names used for different frequency bands and the requisite wavelengths.

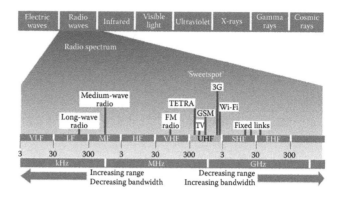

Figure 3.1   The electromagnetic spectrum. (From The Communications Market 2005, Ofcom, http://stakeholders.ofcom.org.uk/binaries/research/cmr/comms_mkt_report05.pdf)

Table 3.1   Names, Frequencies and Wavelength Relationships

BAND	FREQUENCY	WAVELENGTH
Very low frequencies (VLFs)	3000–30,000 Hz (3–30 kHz)	10,000–100,000 m
Low frequencies (LFs)	30–300 kHz	1000–10,000 m
Medium frequencies (MFs)	300–3000 kHz	100–1000 m
High frequencies (HFs)	3000–30,000 kHz (3–30 MHz)	10–100 m
Very high frequencies (VHFs)	30–300 MHz	1–10 m
Ultra high frequencies (UHFs)	300–3000 MHz	0.1–1 m
Super high frequencies (SHFs)	3000–30,000 MHz (3–30 GHz)	0.01–0.1 m (10–100 cm)
Extra high frequencies (EHFs)	30–300 GHz	1–10 cm
Tremendously high frequencies (THFs)	300–3000 GHz	0.1–1 cm

It should be noted that each spectrum band represents a factor of 10 difference in frequency. While this is largely for convenience, the boundaries between different frequency bands have been chosen such that the characteristics within a band are similar, and equally that the characteristics of different bands may be different. It also means that there is much more spectrum in the higher frequency bands than in the lower frequency ones: the super high-frequency band represents a total of 27 GHz, or 27,000 MHz of spectrum and the high frequency (HF) band represents a total of just 27 MHz of spectrum, 1000 times less.

Coverage

As a radio wave travels away from the point at which it is generated, it spreads out, just as waves generated by a stone being thrown in a body of water do, but radio waves spread out in three dimensions, not two. At any given distance from the source, the total amount of energy will be the same (assuming that the wave has not been diverted from its course or absorbed by something). However, as it travels farther from its source, the area over which the energy will have been spread will increase. Imagine a balloon with hardly any air in it. If we were to draw a square on the surface and then blow the balloon up, the square would become larger; however, the amount of material in the square would be the same. So the total amount of material remains the same, but it is spread more thinly. The same is true of radio signals as they spread out through space (Figure 3.2).

Thinking back to the 'stone in the water' scenario, as the waves spread out, the circle they form becomes larger. Each circle has the same amount of energy, but as the circles become larger, the amount by which the water is displaced vertically to form the circle becomes smaller. Eventually, the waves will have spread out so far that the movement of the water up and down will be imperceptible (though with sensitive enough measuring equipment we may still be able to detect the movement).

This holds true for radio waves; however, there is one additional twist. The extent to which the wave spreads out is determined by the number of wavelengths away from the source that we travel, rather than just the pure distance. As such, a wave with a smaller

Figure 3.2   Ripples spreading out over water. (Courtesy of Roger McLassus.)

wavelength (higher frequency) will have spread out much more for a fixed distance from the source than a signal with a larger wavelength (lower frequency). If we move a fixed distance away from the source, we would find that the larger wavelength waves had spread out much less than the smaller wavelength ones. Thus a larger wavelength (lower frequency) wave will appear bigger (stronger) than a smaller wavelength (higher frequency) wave.

Here we have one of the first characteristics of radio waves that might make the value of one part of the spectrum different from another: higher frequency waves spread out and become smaller (or weaker) more quickly over a fixed distance than lower frequency waves. Put it another way, higher frequency waves will become imperceptibly small (i.e. too small to be of use in communicating) in a much shorter distance than lower frequency waves. This fact is often erroneously stated as meaning that lower frequencies travel farther than higher frequencies, which is not strictly true as any radio wave will continue forever, spreading out as it goes, but is nonetheless another way of expressing the same concept. In a similar vein, it might be useful to understand this fact by expressing it as 'the coverage of a lower-frequency radio wave will be greater than that of a higher-frequency wave'.

As an example, let us consider the difference in coverage of two mobile networks, one operating at 700 MHz and one at 2500 MHz. A 700 MHz wave has a wavelength of 43 cm, whereas the 2500 MHz wave has a wavelength of 12 cm. The 2500 MHz wave will become smaller every time it travels 12 cm in distance, whereas the 700 MHz wave will become smaller by the same amount only every 43 cm travelled. The 700 MHz wave will travel 3.6 times (43/12) farther than the 2500 MHz wave before it has spread out (diminished) by the same amount. The coverage of a 700 MHz cell site will therefore be almost 13 times greater (3.6 squared) than that of a 2500 MHz site to the point where the size of the wave has reduced by the same amount. A mobile network at 2500 MHz may therefore require 13 times more cell sites to achieve the same coverage as a network operating at 700 MHz, all other factors being equal (which as we will see, they may not necessarily be).

Thinking back to the example of the wave caused by a stone thrown into a body of water, what eventually causes the waves to become imperceptible is when the waves' perturbations of the water are smaller than the natural ripples on the surface caused by other things affecting the water's surface, such as wind, the pull of the moon and any animals swimming on or beneath its surface. The same is true of a radio wave – it will be detectable (receivable) until such point as it has spread out to the extent that its 'perturbations' of the electromagnetic spectrum are smaller than other 'ripples' such as may be caused by noise and interference, which are discussed in more detail later.

Hitting a Brick Wall

Being part electrical in nature, it is perhaps no surprise to learn that radio waves are drastically affected by things which conduct electricity, particularly metal. In general, radio waves cannot pass through metal. This was discovered by Michael Faraday who is credited with the invention, in 1836, of the Faraday Cage – a metal box which protects any devices enclosed within it from electromagnetic waves and which stops electromagnetic signals generated within it from escaping (as often used by hackers in Hollywood movies to protect their activities from snooping). For other objects, the extent to which they will block, or reflect, radio signals depends on their conductivity and, maybe obviously, their size. But as with the spreading out of signals, it is their size relative to the wavelength of the radio wave which determines how big an effect they have. Thus for a given object, it will tend to impact higher frequencies (shorter wavelengths) more than lower frequencies (longer wavelengths). While this is not a hard and fast rule, it holds generally true, and if we consider, for example, the ability of a radio wave to penetrate a building, it is generally more difficult to get a higher frequency to traverse unaffected from the outside to the inside (or vice versa) than it is to get a lower frequency to do the same. Even tiny obstacles such as raindrops or individual atoms have the potential to impact radio waves if there are enough of them.

Table 3.2 shows the penetration loss for buildings (e.g. the loss caused by radio waves penetrating a building) at different frequencies.

Table 3.2 Penetration Loss at Different Frequencies

FREQUENCY	900 MHz	2000 MHz	5800 MHz
Penetration loss	77% (6.4 dB)	90% (10.3 dB)	96% (13.8 dB)

Source: Interpretation of University of Colorado analysis.

The same data set shows that at 900 MHz, only 23 per cent of a radio wave outside a building will be present inside the building (and vice versa). At 5800 MHz, only 4 per cent of the signal outside is present inside (Table 3.2).

Note that the planet Earth is in itself a large obstacle to radio waves. Most radio waves are unable to penetrate the surface of the Earth (whether land or ocean) and therefore will stop spreading (propagating) at the point where the Earth obstructs them. The distance at which this occurs is known as the *line of sight* and depends on the height of the transmitter (and of the receiver).

Transmitting Signals

To generate radio waves, we use a device known as an antenna (or aerial) whose purpose is to turn an electrical signal into an electromagnetic wave (or vice versa). A long discussion on the properties of antennas is not necessary here, but it is important to know that in order to be able to generate (or detect) a radio wave, an antenna typically needs to be at least a tenth of a wavelength in size, but ideally nearer to a quarter of a wavelength. Thus for a 700 MHz wave, an antenna would ideally be 11 cm in size; for a 2500 MHz wave, an antenna would need only to be 3 cm in size. If an antenna is below its ideal size, it will be less effective in its ability to generate a radio wave. If an antenna is larger than a quarter of a wavelength, it will generally be able to generate a radio wave more effectively. This causes two factors of which we need to be aware:

- For lower frequencies (longer wavelengths), antennas become necessarily larger. At a frequency of 30 MHz, an antenna would have to be at least 1 m in size and ideally 2.5 m. Such an antenna could not, for example, be fitted into a typical mobile device.
- For a fixed-size antenna, it will be more effective at higher frequencies (shorter wavelengths) than at lower frequencies.

To transmit a signal, we need to generate an electrical waveform and connect this to an antenna. The electrical waveform is generated by a device which takes the information that we wish to send (whether analogue or digital) and converts this into a signal at a radio frequency. This conversion is known as *modulation*. The waveform will determine

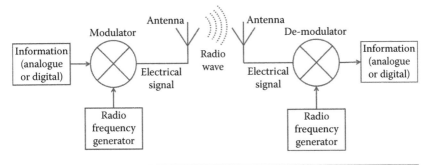

Figure 3.3   Radio transmitter and receiver.

the frequency of operation and the amount of spectrum that the signal will occupy. It will also, to some extent, determine how large the resulting radio wave is; however, this will also be impacted by the effectiveness of the antenna (Figure 3.3).

Different types of modulation are used in different situations. There is a trade-off between the amount of spectrum that is needed, the amount of data that can be sent and the robustness of the signal (i.e. the better its ability to be received). In general, the more power or the more spectrum that is used for the same amount of information, the more robust the signal is.

Receiving a Signal

To receive a signal, we reverse the transmission process. First the radio wave is converted back into an electrical signal by an antenna. The electrical signal is then sent through a de-modulator which recovers the original information. By the time the transmitted wave has reached the receiver, it will have changed in a number of ways:

1. It will have gotten much smaller, as it spread out from the transmitter.
2. It may have encountered obstacles which have prevented all of the wave from reaching the receiver.
3. Some of the wave may have been absorbed (e.g. passing through the wall of a building).

The limiting factor in being able to correctly de-modulate a radio signal to recover the information it is carrying is determined by the point at which it is no longer possible to distinguish the signal from noise. Noise is generated by many natural phenomena such as the sun, the weather (e.g. lightning storms) and even the creation of the universe, but it is also generated by electrical devices (somewhat ironically referred to as 'man-made' noise) including the electronic devices in the receiver itself. Noise is therefore unavoidable. Electronics has made enormous steps forward in overcoming noise, but even today, a typical receiver will generate around the same magnitude of noise as the 'natural' noise, which it encounters in the radio environment it inhabits. Radio astronomers have designed receivers which generate almost zero additional noise, but their sensitivity is still limited by natural noise (Figure 3.4).

Figure 3.4 shows the levels of noise found at different radio frequencies as caused by different sources. Lower frequencies encounter

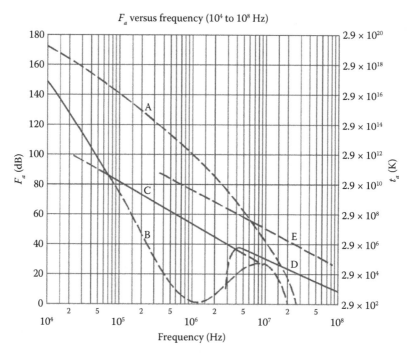

F_a versus frequency (10^4 to 10^8 Hz)

A : atmospheric noise, value exceeded 0.5% of time
B : atmospheric noise, value exceeded 99.5% of time
C : man-made noise, quiet receiving site
D : galactic noise
E : median city area man-made noise
——— minimum noise level expected

Figure 3.4   Graph of noise at different frequencies. (From Recommendation ITU-R, Radio Noise, p. 372–11. http://www.itu.int/dms_pubrec/itu-r/rec/p/R-REC-P.372-11-201309-I!!PDF-E.pdf)

Table 3.3   Characteristics of the Radio Spectrum

CHARACTERISTIC	IMPACT
Amount of spectrum available	There is proportionally more spectrum in higher frequency bands than in lower frequency ones.
Reduction of signal with distance	Higher frequencies spread out more than low frequencies (and hence on a like-for-like basis higher frequencies will provide smaller coverage).
Penetration of obstructions (e.g. buildings)	Lower frequencies are more able to penetrate obstacles than are higher frequencies.
Size of antennas	Higher frequencies need smaller antennas than do lower frequencies for the same performance levels. Alternatively, antenna performance (for a fixed-size antenna) is better at higher frequencies.
Noise	There is less noise at higher frequencies.

much higher levels of noise than higher frequencies. Noise at higher frequencies is typically limited by the electronics in the receiver.

We therefore now have all the technical pieces in place to understand the characteristics of different parts of the radio spectrum, which may have an impact on its value. These are summarised in Table 3.3.

There is one other set of effects which vary with frequency, and that is the extent to which the Earth's atmosphere impacts radio waves. The majority of such effects occur on frequencies below around 300 MHz, though in exceptional conditions, effects can extend to 10 GHz or higher. In general, the impact of the atmosphere is to extend the ability of a signal to travel farther than the normal line of sight. In some cases, this effect is beneficial: it is the mechanism that is the core of short-wave broadcasting which uses the upper layers of the atmosphere (the ionosphere) to extend the range of HF waves from line of sight to 2000 km or more. In other cases, the effects can be detrimental: signals from far away can become as strong as those from local transmitters causing unwanted reception problems.

Interference

In addition to noise impeding reception, there may be other transmitters using the same frequency as the one we are trying to receive. In this case, we need some way of distinguishing the signal that we want from the others that we do not want. The simplest way to do this is by ensuring that the signal generated by the radio wave that we want is stronger than any unwanted signals (and the noise), but there are other methods, too. Unwanted signals are known as interference, and in addition to being caused by other transmitters on the same frequency they could also be caused by transmitters on neighbouring frequencies which are too close to the wanted frequency for us to properly disentangle them.

Transmitters generate a wanted signal (i.e. that which contains the information that has been modulated onto the intended frequency of operation) and, as a result of the practical limitations of the technology used, also produce unwanted, spurious outputs on other frequencies. Receivers have similar spurious responses and in particular find it very difficult to distinguish signals on the wanted frequency from those on frequencies immediately adjacent to it. The ability of a receiver (and a transmitter) to focus on the wanted signal as opposed to any spurious signals is determined by a device known as a *filter*. The filter may be an analogue component (whose performance is limited by its size and complexity) or may be implemented digitally (whose performance is limited by the power of the processor undertaking the complex mathematics involved). Historically, all radio equipment employed some form of analogue filter somewhere in its design. More recently, radio equipment employing purely digital techniques has been developed. Known as 'software radios', they conduct all the modulation and de-modulation in software and can implement filters with much better performance than their analogue counterparts. There are still limitations in performance caused by the digital hardware and the processors, but as both these elements improve with time, some of the limitations in the use of spectrum caused by the problems of imperfect filters may eventually cease to be such concrete restrictions.

It is also worth noting that a transmitter in close physical and spectral proximity to a receiver will almost inevitably cause interference. In a mobile handset (or base station), for example, the transmitter sending information in one direction is in the same physical space and uses the same antenna as the receiver. In order to stop the transmitter from interfering with the receiver, filters are used to separate the two. These filters are not perfect and cannot separate two immediately adjacent frequencies so a gap is left in frequency between that used for the transmitter and that used by the receiver. This is a reason why many two-way radio systems (where a user can both talk and listen simultaneously) use two pieces of spectrum, often referred to as paired spectrum, which are far enough separated that the transmitter and receiver can operate effectively within the constraints of the filter's performance.

Spectrum Efficiency

There is a direct trade-off between the amount of spectrum required to transmit information, the amount of signal available at the receiver and the robustness of the signal to noise and interference. For the same degree of robustness, the same information could be carried using more spectrum with less power, or less spectrum but with more power. The amount of spectrum used to transmit a certain amount of information is one measure of spectrum efficiency. The number of bits per second of data that can successfully be transmitted in each Hertz of spectrum is a metric often used to compare one system with another (Table 3.4).

For a device such as a mobile phone, where the amount of power is restricted by the battery life, there is a trade-off between the amount of spectrum used and the robustness of the signal. It is perfectly possible to use far more spectrum than is strictly needed to transfer the information and in return make a very robust signal, one that is capable of being received even if it is weaker than noise or other

Table 3.4 Examples of Spectrum Efficiency of Different Technologies

TECHNOLOGY	EFFICIENCY (BITS PER SECOND PER HERTZ)
GSM[a] (2G)	1.3
UMTS[b] (3G)	2.4
LTE[c] (4G)	4
DVB-T (digital terrestrial TV)	2–4
DVB-S2 (digital satellite TV)	2–4

[a] Second-generation digital standard for cellular mobile phones.
[b] The Universal Mobile Telecommunications System (UMTS) is a third-generation mobile cellular system based on the GSM standard.
[c] Long Term Evolution (LTE) is often referred to as 4G. This enhances the GSM and UMTS technologies to offer higher data rates and other improvements.

interfering signals. This technique of stretching a signal out across more spectrum than is absolutely necessary is called *spread spectrum*, and is a common technique as it allows multiple transmitters to use the same frequency and interfere with each other while still allowing the receiver to decode all of them successfully.

It would be possible to vastly increase the robustness and thus the range of the signal from a mobile cell by increasing the power of the transmitter or by using much more spectrum, but this is not efficient in terms of the use of spectrum or the use of energy, and there are therefore practical limitations as to what is realistically achievable.

In the late 1940s, Claude Shannon and Ralph Hartley produced a simple equation which allows us to determine the maximum capacity of any communication channel (including radio channels). Their theorem tells us that the amount of information that can be carried in any channel is directly related to the strength of the received signal compared to any noise or interference (known as the signal-to-noise ratio), and the amount of spectrum used. Engineers often use the Shannon–Hartley theorem to assess how effective a particular technology is at using the radio spectrum.

Systems which operate in environments that are relatively fixed in nature (e.g. fixed satellites where the path between the transmitter and receiver varies very little with time) can achieve performance much closer to the Shannon limit than those which have to deal with wildly varying conditions (e.g. mobile systems). In the case of mobile signals, as the mobile user moves around, the path between them and the cell site will constantly change. The obstacles between the two will change, and other objects that might change the strength of the signal (such as reflections off buildings or vehicles) will also not be constant. This causes the signal to vary in size in an unpredictable way. Mobile technology attempts to compensate for this by continuously varying the power of the transmitter to try and keep the received signal constant, but there is still a need to add in a 'safety margin' to ensure that variations not able to be dealt with by the power control algorithms can be accommodated without loss of communication. This safety margin is effectively wasting more power (or maybe spectrum) than is strictly necessary to maintain the connection and therefore reduces the overall efficiency of the system.

Table 3.5 Ideal Spectrum for Different Radio Services

SYSTEM	IDEAL SPECTRUM	REASON	EXAMPLE
Radio time signals	VLF/LF	Very large coverage is appropriate. There are few restrictions on the size of the transmitting antenna. Not much spectrum is required to carry the information.	The German time station (DCF) uses 77 kHz. The British time station (MSF) uses 60 kHz.
High-quality audio broadcasting	VHF	Good coverage characteristics are required. Transmitting antennas can be relatively large – receiving antennas can be reasonably large. A reasonable amount of spectrum is required to carry the information.	FM broadcasting uses frequencies around 100 MHz (3 m).
Mobile phones	UHF	Coverage is important (though the cellular concept relies on being able to reuse frequencies time and again meaning coverage is not the ultimate goal). Transmitting antennas can be large; however, receiving antennas (on mobile handsets) need to be small. Significant amounts of spectrum are required to carry the information.	Mobile networks typically operate on frequencies in the range of 700 to 2600 MHz. Sophisticated technology and the use of smaller cells may extend the frequencies that mobile operators use.
Satellite broadcasting	SHF	There are few issues with coverage as communication is line of sight. Lots of spectrum is needed to carry the information.	Satellite broadcasting operates in the frequency range of 3400–12,750 MHz (3.4–12.75 GHz).
Point-to-point links	SHF	Communication is line of sight so coverage is not an issue. Lots of spectrum is needed to carry the information.	Fixed links typically operate in the frequency range 7–40 GHz and sometimes even higher in frequency.

The characteristics of different radio uses and services naturally lend themselves to certain pieces of radio spectrum. Table 3.5 provides examples of the ideal part of the spectrum for different radio services.

In the next section of this book, we consider how theories about spectrum liberalisation have been applied in practice, looking in turn at auctions, licensing, public sector spectrum, broadcasting, satellite, ultra wide band, Wi-Fi and TV whitespace.

PART II
LIBERALISATION IN ACTION

4

SPECTRUM AUCTIONS BECOME THE DEFAULT ASSIGNMENT MECHANISM

The success of auctions was a boost for the spectrum liberalisers' worldview. They argued that spectrum should be seen as a commercial input like any other, and this could be distributed in the same way as other scarce resources, through a price mechanism.*

An auction is the price mechanism in its simplest form: whoever pays the most wins. As discussed in Chapter 2, the huge sums paid at auction in the United States and in Europe from 1995 to 2000 confirmed the benefits of treating spectrum as a commercial input and encouraged other countries to take this approach.

Auctions have been much more successful than their predecessor, beauty contests (which are described in more detail below), as the default mechanism for assigning spectrum. Beauty contests, or administrative assignments, as they are more properly known, take too long and are often subject to legal challenge.

Problems with Beauty Contests

A beauty contest is the jokey term for a tendering process where competing bodies write bids in which they seek to persuade regulators that they should receive a given block of spectrum. Regulators then judge these bids according to a set of criteria that reflect whatever policy goals a regulator might have (which may include the price that bidders are prepared to pay). By contrast, in an auction, the price that bidders are willing to pay is the only criterion that is considered.

---

* A paraphrase of Coase, R. H. (1959). The Federal Communications Commission. *Journal of Law and Economics*, Vol. 2, p. 14.

Beauty contests have three main failings: they are slow, they are insufficiently transparent and therefore vulnerable to legal action, and there is an incentive for participants to overpromise and underdeliver.

A Slow Process

Typically, companies participating in beauty contests for mobile licences are judged on criteria such as the following:

- Financial strength of the company
- Technical plan
- Business plan
- Experience and knowledge
- Geographical coverage
- Roll out timescale

The price that the bidder is willing to pay may also form an assessment criterion.

Putting all this information together creates a very long document: in the United States, the Federal Communications Commission (FCC) said many of its beauty contest applications were over 1000 pages long. For the first award of mobile licences in 1981, it received 200 applications, and it seems reasonable to estimate that just reading these documents would take one person about a year, or a team of six about 2 months. Comparing and analysing them would take considerably longer time.

'The task of evaluating and then awarding the licenses in an informed and equitable manner put a strain on Commission resources', said the FCC with considerable understatement. 'The selection of licensees from a pool of applicants often took up to two years or longer to complete'.[*]

Beauty contests certainly consume regulatory resources but the more important issue is that they delay the process of getting new services to market. Every year wasted is a year when the wider economy fails to benefit from the efficiencies and new products that enhanced mobile services can provide. One study argued that cellular licensing in the United States had been delayed by 10 years in total

---

* Federal Communications Commission, Report to Congress on Spectrum Auctions, 1997, p. 7.

and this had cost the US economy the equivalent of 2 per cent of gross national product.*

The snail's pace of the beauty contest is a powerful argument for moving to auctions, but in 1982 the FCC decided to use lotteries for assigning licences, which did little to speed things up and created more serious problems. The idea was for the regulator to compare the applications in a less rigorous and less time-consuming manner, assessing whether applicants passed a basic level of suitability, and from this group of qualified companies, deciding the winners in a process similar to drawing names from a hat.

Even this prescreening process was taking up to 20 months, so in 1987 the FCC was forced to abandon it and open the process to anyone. As it cost $800 to file an application for a licence worth millions, something akin to a gold rush ensued. Approximately 400,000 firms tried to win licences and 'application mills' sprang up to provide the relevant paperwork.†

One licence was even won by a group of dentists who sold it to Southwestern Bell for $41 million.‡ It was 'unjust enrichment': people with no knowledge of the mobile industry were making money by selling licenses on to companies that did. Surely Southwestern Bell should be paying $41 million to the US government, not to a bunch of dentists.

Even worse, the process of selling on the licences to companies that could actually operate mobile networks took several years. Lotteries were proving no quicker than beauty contests, and in 1993 the FCC started the process of finding a suitable way of selling spectrum by auction.

The advantage of auctions is that they are much quicker. The consultations on the auction format may take some time, but so would a consultation on the criteria to be used in a beauty contest. Once the format has been agreed, the auction can be over in a few days or less: there is no need for regulators to spend months or years comparing lengthy documents. It is true that some auctions for the sale of hundreds of regional licences in the United States and in Canada have gone to hundreds of rounds, but this is very much the exception.

---

* FCC Report to Congress on Spectrum Auctions, 1997, p. 8.
† FCC Report to Congress on Spectrum Auctions, p. 7.
‡ *The Bidding Game.* (2003). *Beyond Discovery* published by US National Academy of Sciences http://www.nasonline.org/publications/beyond-discovery/the-bidding-game.pdf

Legal Challenges

It takes a long time to compare the suitability of different applicants in a beauty contest, but a further problem is the inherent subjectivity of many of the criteria mentioned above. If it is not crystal clear why one applicant has lost and another has won, then these are fertile grounds for legal challenge.

Business plans are necessarily about predicting the future: How can the regulator – or anybody else – prove that the applicant's claim of getting a 25 per cent market share is unachievable?

What experience or knowledge is required to run a successful mobile network? A regulator may dismiss an application because it lacks technical expertise, but the rejected company could perfectly well argue that mobile is primarily a business which requires strategic vision and sufficient funding. Technical knowledge can be brought in at a later stage.

One famous legal challenge was brought by Orange against the 1998 award of Ireland's third mobile to a company called Meteor. Orange claimed as a UK company it had been discriminated against in favour of Meteor, which was an Irish/US company. Its claim was upheld in Ireland's High Court but eventually rejected in the Supreme Court in May 2000. Chief Justice Ronan Keane said Orange's claim that it has suffered xenophobic prejudice 'strained credulity to breaking point'.* His ruling continued: '[T]he fact that a particular bidder was better acquainted with the circumstances of the Irish market and had sought to tailor the bid somewhat more closely to Irish conditions than the other bidder was a factor that the [regulator] was perfectly entitled to take into account', he said.

Meteor suffered a 2-year delay in receiving its licence, which it estimated had cost the company £100 million.

There was also a legal challenge when Sweden awarded its 3G licences by beauty contest in 2000. The licences went to two incumbent Global System for Mobile (GSM) operators Europolitan Vodafone and Tele 2, as well as new entrants HI3G and Orange Sverige. One GSM incumbent, Telia, which was partly owned by the government, did not receive a licence, which came as a great surprise. Telia, along

---

* Raidió Teilifís Éireann (RTE) News, 18 May 2000. Meteor to enter mobile market within nine months. http://www.rte.ie/news/2000/0518/6995-business1/

with fellow losers, Reach Out Mobile and Telenordia, launched a court action, but this was rejected.*

Beauty contests have developed a reputation as the assignment method which can land the regulator in the courts. No official wants to waste time justifying previous decisions in front of a court instead of tackling the problems of today and tomorrow, so auctions have become the more popular method of apportioning the airwaves.

There are countries where beauty contests continue to be used in favour of auctions. In Malaysia, for example, the regulator continues to prefer beauty contests as the perceived propensity for gambling in the country could lead to auction outcomes that are more based on bravado than a desire to operate a mobile network.

Auctions are rarely challenged in the courts because the criteria for winning or losing are very clear: the fact that one bidder has paid more than another is not open to debate, unlike the more nebulous criteria used in beauty contests such as the suitability of the management team.

An Incentive to Overpromise

In the Swedish beauty contest described above, one of the criteria for choosing the winners was geographical coverage to fulfil a policy goal of making 3G available in 99 per cent of the country's surface area by the end of 2003.† But by 2005 the Swedish regulator, the Swedish Post and Telecom Agency (PTS), was complaining that operators were not fulfilling the commitments made to win the licences. The PTS said it was working 'to urge the operators to roll out their networks to the extent that they have promised'.‡

Here we see a major problem with beauty contests: to win there is an incentive to promise more than the other participants, but once you have won there is an incentive not to deliver because these promises of extra coverage or higher speeds involve higher costs.

---

* Björkdahl, J., and Bohlin, E. (2001). *Financial Analysis of the Swedish 3G Licensees, Where are the profits?*, Vol. 4(4), pp. 10–16.

† Ibid., p. 6.

‡ *Sweden has the best 3G coverage in Europe*. PTS press release. 22 February 2005. http://www.pts.se/en-GB/News/Press-releases/2005/Sweden-has-the-best-3G-coverage-in-Europe/

It is difficult for the regulator to withdraw the licence because of a failure to deliver. Customers could lose their service, and terminating a licence would make the market less competitive and less attractive to investors. Furthermore, the operators' promises may not be legally enforceable.

There is an inherent moral hazard with beauty contests: regulators make a decision but they do not suffer the consequences of that decision. It is the customer who will suffer through poorer services or lack of competition.

To contrast this with auctions, if operators offer a large sum of money to outbid their competitors, then they *do* have to face the consequences of their actions. They must be certain that they can win back the amount of money they have committed, so their willingness to pay a larger amount should be based on a thorough assessment of the market. There is no incentive to overpromise: bidding too much could leave the winner facing bankruptcy.

In fact, the large sums paid during spectrum auctions are an incentive to roll out networks. Operators will either need to pay back the money they have borrowed or show some return to shareholders. The logical way to do this is to make use of the spectrum by providing new services.

The Market Does the Regulators' Work

Perhaps the most persuasive argument for auctions is that they also fulfil the comparative function carried out by regulators in a beauty contest, and may even do a better job. Typically, in beauty contests a regulatory team of upwards of five people has to decide which market projections are credible, which plans are technically viable and which company has a competent management team – in auctions the market does this work.

In auctions operators have to convince shareholders, bankers and/or the stock market of the wisdom of their potential spectrum purchase. These bodies have more resources than regulators and will undertake due diligence involving the best accountants and lawyers. If they doubt the business case or the competency of management, then the operators will not get the money to buy the licence.

So in an auction, the market carries out its own testing of operator's business plans, just as regulators do in a beauty contest. This is how an auction should work in theory, but as it is often pointed out, the overpayment during the 3G auctions in Europe in 2000 shows this does not always happen in practice. However, this sort of overvaluation is relatively rare, and we discuss this in more detail in Chapter 18.

Auctions Dominate

By the second decade of the twenty-first century, auctions had become the default assignment mechanism for high-value spectrum – particularly mobile licences. The idea that mobile spectrum is a valuable state asset and it is fairest and most efficient to sell this to the highest bidder has been accepted almost universally.

In the spectrum field, auctions are the most widely used market mechanism, but they do not necessarily indicate support for the whole liberalisation agenda. Many countries who hold auctions would not contemplate trading or pricing public sector spectrum.

There is another reason why spectrum auctions have become so popular: they are a huge windfall for the treasury, something which any government finds hard to refuse. There are growing concerns that the benefits of this cash injection may be outweighed by the wider social cost of delayed network investment, an argument which we examine in Chapter 18. The worry here is that governments have become addicted to the short-term financial benefits of selling spectrum, while failing to recognise its wider economic significance as a means for delivering universal broadband. Some critics of liberalisation see the international enthusiasm for one narrow aspect of it – spectrum auctions – as being one of the reasons for its failings.

5

The Liberalised Licensing Debate

Introduction

In order to control spectrum usage, regulatory authorities issue licences in the same way that road usage is controlled by transport authorities issuing driving licences. These licences specify the terms and conditions under which the licensee can use the radio spectrum. Such licences are necessary to ensure that users operate responsibly and thereby maximise access to the spectrum. Licences can be issued in a variety of ways from command and control through beauty contests and auctions. They can even be bought and sold through trading, where this is enabled.

Technology-neutral licensing has been one of the most successful and widely adopted aspects of spectrum liberalisation, largely replacing the previous approach where licences required users to deploy a specific technology. The neutral approach allows users to upgrade their networks as superior technologies become available, meeting some of the key goals of spectrum liberalisation, namely, improving efficiency and allowing an easy progression to higher value uses of the airwaves.

In this chapter, we discuss how technology-neutral licensing works in a technical sense and examine the possibility of using licensing regimes which give even greater flexibility in spectrum use.

Defining Spectrum Use

Any radio transmission will use a piece of the radio spectrum. However, defining a piece of spectrum requires more than just a reference to a particular radio frequency. Spectrum can be reused in different geographic areas and at different times. A piece of spectrum can therefore be defined by a number of distinct parameters:

- *The frequencies in use*. This is, in principle, straightforward and is defined by a lower boundary and an upper boundary.

- *A geographic area.* The area occupied by the service may be defined as the area of which it is expected that a service can be provided (the service area), over an area where the service is protected from interference (the protected area) or the area over which the service would preclude the use of the frequency by other users (the area sterilised).
- *A time period.* Normally this is 24/7; however, it is possible to licence a service for a restricted amount of time, whether it be a portion of each day or certain days. Many licences have an expiry date and thus are time bound in this respect.

One of the difficulties in defining spectrum use is that radio transmissions do not suddenly stop when they reach a boundary, whether that boundary is in geography, time or frequency. The definition of spectrum use therefore needs to take account of any 'spillover'. In general, such spillover will have the effect of causing signals to appear in neighbouring pieces of spectrum, whether neighbours in geography or in frequency. If these signals exceed acceptable levels, they will disturb other users in neighbouring pieces of spectrum potentially causing their services to be interrupted. These interruptions and disturbances are termed *interference*. There are different degrees of interference depending on the extent to which a service is affected. The highest level of interference is known as *harmful interference* and is defined by the ITU as follows:

> Interference which endangers the functioning of a radionavigation service or of other safety services or seriously degrades, obstructs, or repeatedly interrupts a radiocommunication service operating in accordance with these Regulations.

It is therefore important to define not just the piece of spectrum that the service is licensed to use, but also the extent to which it is permitted to cause (harmful) interference to neighbouring users.

Traditional Licensing

Historically, and typically under command-and-control licensing regimes, when issuing a licence permitting the use of the radio spectrum, regulators would define a series of administrative and technical parameters to which the licensee must adhere. Such parameters would form proxies for defining the piece of spectrum in use and may comprise elements such as the following:

- *Transmitter power.* Changing the power of a transmitter will impact both its geographic coverage and, potentially, the amount of interference it causes to its neighbours in the spectrum.
- *Antenna height and directivity.* Increasing the height of an antenna will increase its geographic coverage. Using a directional antenna focuses transmissions into a specific area and therefore changes coverage.
- *Frequency of operation.* This is typically a centre frequency, about which the service is permitted to operate.
- *Technology permitted (or modulation scheme).* Defining a technology specifies how much spectrum can be used (e.g. defining stereo FM broadcasting as a technology implies spectrum use totalling just over 250 kHz, whereas defining AM broadcasting implies spectrum use of just 10 kHz). Defining a technology also largely defines the extent to which a transmitter will generate interference to its neighbours in the spectrum.
- *Transmitter location.* This together with parameters such as power and antenna height define the area over which the service can operate.
- *Receiver location.* For some services (such as point-to-point links), the receiver location is needed to define the area over which the service uses the spectrum.
- *Field strength.* Some regulators define the field strength that can exist at the boundary of the service, leaving the licensee to determine how to achieve this (e.g. what power and antennas to use). Normally, this was only used for low-power services with limited coverage areas (such as the Part 15 rules in the United States).

For services where usage is unlikely to change over long periods of time, such parameters are a sound and logical way to define spectrum use. However, these parameters constrict the ability of the licensee to use the spectrum in a flexible way, or to join together multiple pieces of spectrum and use those pieces in a different way.

As each licence is highly restrictive in the latitude, it gives the licensee to change any element of the use of the spectrum, it would not be possible for an organisation to gather together multiple licences covering adjacent geographic areas and then repurpose the spectrum across the wider area, even if the change would have no impact on any of the neighbours. In addition, the process for issuing licences using such parameters can be very long winded, especially if it involves the introduction of a new technology. Lengthy compatibility studies which assess the likelihood and extent of interference between neighbouring users are usually conducted before any new usage is permitted.

Technology-Neutral Licences

With demand for the radio spectrum increasing, and new technologies coming to market at an ever-increasing rate, several regulators have considered methods for licensing spectrum that make its use more flexible and permit licences to be traded in a meaningful way such that, for example, geographically adjacent licences can be amalgamated. The first stage in this liberalisation process is the removal of the definition of a technology in the licence, leading to technology-neutral licences.

In reality, defining spectrum use by specifying a technology is actually defining two characteristics of the transmitters used:

- The band of spectrum occupied by the wanted part of the transmission – the 'in-band' emissions
- The extent to which the signal spills out into adjacent frequencies – the 'out-of-band' emissions

As long as a replacement technology follows the same in-band and out-of-band emission patterns, changing the technology would not fundamentally change the way in which the spectrum is used, or the interference to neighbours. It is perfectly feasible to define in-band and out-of-band emissions without reference to a specific technology, and this is done through the use of 'masks'. Masks commonly come in two types, a block-edge mask which defines the levels of in-band and out-of-band emissions for a particular block of spectrum (often a single transmission) and a band-edge mask which defines the levels

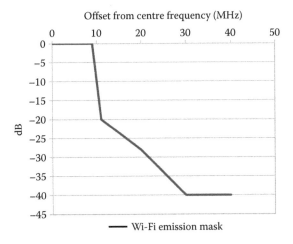

Figure 5.1   Example emission mask.

of emissions over a whole band of frequencies. Both are similar in appearance, and an example mask is shown in Figure 5.1.

In this instance, the mask shows the relative level of emissions permitted within the licensed spectrum (0 dB) and the emissions permitted outside the licensed spectrum (in this case, 40 dB lower at frequencies of more than 30 MHz away) but does not specify technology. That being said, most masks are based on a particular technology. So, for example, a block-edge mask for a digital television transmitter will normally be based on the performance of the self-same transmitter. While this may seem counterproductive insofar as it just replicates the parameters defined in the technology version of the licence, it does allow the licensee to change technology, as long as the new technology complies with the same limits originally set (in this case, digital broadcasting) and thus is neutral to changes in the technology.

Attempts have been made to define masks that are not just based on one technology. To do this requires that the limits are set to the worst possible case emissions of the technologies being considered such that each technology will fit within the mask. One of the problems of such an approach is the widely varying differences between the in-band and out-of-band emissions produced by different technologies. Figure 5.2 shows the masks for a wide range of different technologies.

Though this may seem a confusing diagram, the key point is the fact that the masks are all very different from each other and that the

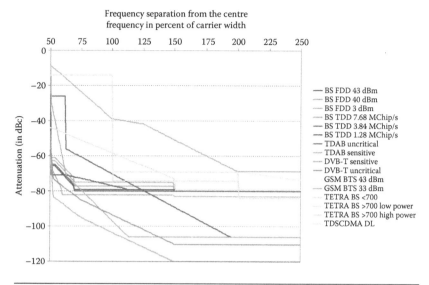

Figure 5.2 Emissions masks for multiple technologies. (From Study on Radio Interference Regulatory Models in the European Community, *Eurostrategies/LS telcom*, 2007, http://bookshop.europa.eu/en/study-on-radio-interference-regulatory-models-in-the-european-community-pbKK0414494/)

choice of one mask would not be applicable to a different technology. Further, if a mask were designed to accommodate the worst-case performance of all of these systems, it would permit such high levels of interference that the majority of them would not function. This is not an efficient way to manage the radio spectrum.

The European Commission established the Wireless Access Platform for Electronic Communication Services (WAPECS) policy to try and allow all current and foreseeable future technologies to be used in the main mobile bands. In doing so, it has succeeded in defining a set of 'least restrictive technical conditions' which are applicable to multiple technologies but which do not simply do this by increasing the permissible out-of-band interference levels. By considering the mobile technologies available at the time, it defined a mask which can be equally applied to Universal Mobile Telecommunications System (UMTS) and WiMAX technologies. By licensing mobile network operators using this mask, it would therefore be possible for them to migrate from one technology to another within their own spectrum without the need for a licence modification to permit the change in technology.

Those developing the standards for LTE seized upon this idea and ensured that the in-band and most importantly the out-of-band emissions produced by an LTE transmitter also remained within the same mask. Thus any licensee whose licence was based on the WAPECS mask could also re-farm their spectrum and introduce LTE services in spectrum that was previously used for Global System for Mobile (GSM) or UMTS without breaching the interference limits set by the mask offering even greater technology flexibility.

Licensing Based on Reception Not Transmission

Some regulators have attempted to move beyond spectrum masks (which are applied to transmitters) to consider not just the emissions caused by a transmitter, but the impact of those emissions on the neighbouring spectrum users. The idea is that it is not the emissions from one transmitter that cause interference, but the combined emissions from multiple transmitters together with the distance between those transmitters and any victim receiver. As such, defining the emissions at the victim receiver should offer a user the maximum flexibility in choice of technology and in the positioning of sites while still offering protection from interference to the victim. In the United Kingdom, Ofcom's spectrum usage right–based licences attempted to use this approach. Only one licence has been using this approach, that for the use of the L-Band (1452–1492 MHz) which was won by Qualcomm.

In principle, applying such a scheme is relatively straightforward. At the location of any victim receiver, the level of incoming interference can be measured. As with any other licence conditions, if the levels were found to be in excess of the set limits, the victim would approach the regulator to conduct enforcement activities. In the case where the victim's receiver is a stand-alone device, such practices would be easy to apply. However, in the case of a mobile network, where the victim receivers are both mobile and also could be located almost anywhere with reference to the interfering transmitters, its application becomes far more complex. If it is assumed that there are two networks operating alongside each other and one suspects interference from the other, the first network would have to turn off its own transmitters in order to conduct measurements of the levels of emissions being caused by its neighbour. This is clearly not practical in a real-world situation.

An alternative way to achieve the same outcome would be to model the emissions of a network of transmitters to predict the level of interference that would be caused to neighbouring users. Again, in the case of a mobile network, where the victim receivers could be anywhere, such an analysis would also have to consider the locations of the victims. Ironically, in order to conduct such predictions, the in-band and out-of-band emission mask for each transmitter would need to be known together with its location, antenna height and transmitter power.

And so we seem to have come full circle. In moving away from a tightly defined licensing regime into a more technology-neutral approach, we have ended up needing to know exactly the type of information that would have been needed for the non-neutral licence approach. While defining emissions at receivers is a noteworthy and interesting concept, its practical application is less straightforward. It may have had some application in the case of LightSquared whose desire to use the spectrum was rejected on the basis of interference into receivers rather than emissions from transmitters, though it is questionable whether agreement could have been reached on the technical characteristics that should be used to represent the receivers.

This is not to say that a technology-neutral approach has no merit. The introduction of masks to define emission limits does provide a means to allow licensees to change the way in which they use their spectrum and offers far more flexibility to use the spectrum in different ways. This approach is gaining traction among regulators especially in bands where usage might foreseeably change. But going beyond technology-neutral licensing to consider not just one transmitter but a whole network is beset with numerous practical and implementational problems, setting a limit on the application of liberalisation to licensing.

6

Public Sector Spectrum

There is a growing risk that spectrum hoarding by the public sector will constrain the growth of private enterprise.

Martin Cave
Review of Radio Spectrum Management, 2002

Public sector spectrum refers to the use of the airwaves by the military, the maritime sector, aeronautical services, the emergency services and other governmental bodies. Perhaps surprisingly, about half of the spectrum below 6 GHz is typically reserved for the public sector.[*]

There is great interest in releasing more of this spectrum for commercial services, but the application of liberalised principles to this sector is particularly difficult. For public sector users, the stakes are far higher than in the commercial sector. For air traffic controllers or police officers, unreliable wireless communication may be literally a matter of life and death; in the commercial sector it is more likely to mean an annoying dropped call which can be redialled a few moments later.

Many public sector services are also dependent on international harmonisation. The safety of planes and shipping depends on them being able to communicate on internationally harmonised channels: the armies of the disparate countries in the North Atlantic Treaty Organization (NATO) – from Canada to Turkey – need to be able to talk to each other when they take joint military action. The international nature of much public spectrum use makes the application of market mechanisms more difficult. In fact, the usage of spectrum in some parts of the public sector is governed not so much by

---

[*] RSPG. (2009). *Opinion on Best Practices Regarding the Use of Spectrum by Some Public Sectors*, p. 3. http://rspg-spectrum.eu/wp-content/uploads/2013/05/rspg09_258_rspgopinion_pus_final1.pdf

national regulators, but by consensus in international institutions such as the International Civil Aviation Organization (ICAO) or the International Maritime Organization.

Several methods have been tried to increase the efficiency of public sector spectrum users, and this chapter focusses on administered incentive pricing (AIP), which has been used in the United Kingdom since 1998.

Uses of Public Sector Spectrum

The public sector's use of spectrum has become a target for liberalisers because it tends to be managed with a different set of considerations to commercial spectrum, and often by a different body. In the United States, for example, the commercial use of spectrum is managed by one body, the Federal Communications Commission (FCC), whereas the use of the spectrum by the federal government is managed by a different body, the National Telecommunications and Information Administration (NTIA). Similar models exist in many other countries. Regulators managing commercial spectrum have its value and therefore its efficient use at front of mind, and with public sector usage the motivations are usually national security and safety of life.

To take the example of radar for transport, in a typical European country 21 per cent of spectrum below 6 GHz is used by maritime and aeronautical radars.* The techniques that are deployed by the world's aeronautical and maritime authorities have scarcely changed for decades. When judged from the metric of public safety, these systems score high – they are extremely reliable and safe. But there is an ongoing debate about the spectral efficiency of these radars.

The other big public sector user of spectrum is the military. In NATO countries, the military uses between 1 and 3.5 GHz for a range of services such as search radars, tactical relay radios, and increasingly, unmanned aerial vehicles – better known as *drones*. Typically around half of this spectrum is shared with other users such as civil aviation. National regulators often have little control of what the military does with spectrum and does not levy

---

* Ibid., p. 3.

substantial fees, even though the same regulators charge private mobile operators millions of pounds for similar frequencies.

Most administrations have taken administrative approaches to compelling the public sector to use spectrum more efficiently.

For example, the Netherlands has adopted a 'justification procedure' for public sector spectrum. Under this system, a public body is obliged to submit a detailed explanation of its spectrum requirements every 3 years. The economics ministry would then use this information to decide whether or not the public body's needs warrant the use of that spectrum. Similarly, in Japan, the Ministry of Internal Affairs and Communications holds periodic surveys of radio spectrum usage and uses this information to decide which services need to vacate which frequencies. This is paid for through a state fund.

In France, public institutions negotiate with each other, and this process can lead to an improvement in spectrum efficiency. For example, the French government has long wanted the 1700–2000 MHz band to be cleared of military use, and the military has sought to gain a bespoke fibre-optic network and a waiver from spectrum fees in order to do this.

Theory of Administered Incentive Pricing

Liberalisers would like to introduce an economic incentive into the assignment of spectrum. Most countries do not follow this principle in the public sector, and the huge amounts of spectrum it occupies have rung warning bells for liberalisers. However, the practical application of this principle to the public sector is difficult.

Many government bodies do not possess private-sector-style licences for the spectrum they use. They just use the bands in which their equipment works. Indeed, they are often underresourced and do not see why they should pay to use an input that does not directly cost the government anything. Audits of public sector spectrum carried out in the United Kingdom and the Netherlands found that not all government bodies knew what spectrum they were using,[*]

---

[*] See *PolicyTracker* Special report, April 2008 (https://www.policytracker.com/headlines/special-report-do-we-know-how-much-spectrum-the-public-sector-uses).

nevermind what its value was. Conducting these audits has usually been the first step for countries wishing to reform public sector spectrum usage, and it is likely that there is even less knowledge of public sector spectrum holdings in many countries that have never started this process. Very often the information is withheld, citing reasons of national security.

One way of introducing a measure of liberalisation to public sector spectrum use is spectrum pricing. The idea is that obliging the public sector to pay for the spectrum it uses – as it does for most other inputs – would mimic the private sector spectrum market, and therefore drive efficiencies in the use of spectrum. Over time, it is hoped that public sector bodies would learn the value of the spectrum it occupies and, once it has to pay for them, sell any superfluous frequencies. Equally, if its needs increased, it could buy other bands on a free market. This idea was proposed by Coase in 1959, as explained in Chapter 2.

There are several ways of setting an incentive price for radio spectrum, including the opportunity cost method proposed by Coase. Opportunity cost is the value of output that is forgone when a frequency is used for one thing rather than another. So if police radios use a band which might otherwise be used for a business radio service like taxis, then the value of that police band is the price that a taxi company might pay for a similar licence. Using opportunity cost as a way of setting an incentive price is known as AIP.

The AIP gives spectrum managers in public sector bodies both carrots and sticks to incentivise them to use spectrum efficiently. The stick comes from the need to pay for all the spectrum they are using, but the body should be given extra money by the government to cover these costs. If the organisation manages to do without certain frequencies, then it can keep the surplus cash and spend it on other activities

Intricacies of Setting Prices

Theoretically, a given frequency band's opportunity cost is discovered when similar spectrum is sold by the government at auction, or between private spectrum licence holders. But some public sector bands have never been sold, and due to the lack of any commercially available equipment, they are unlikely ever to be desired by the private sector. To address this, administrations must make their own

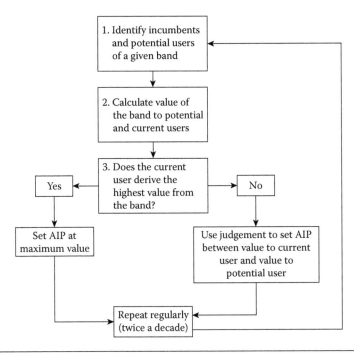

Figure 6.1   Implementing AIP.

calculations regarding the potential value of a band. These calculations lend the word 'administered' to AIP.

In effect, regulators need to establish at what price level a body will forgo its use of a given band. One very influential approach to this was set out in a UK study by the engineering consultants Smith and the economic advisors NERA in 1994.[*] Smith and NERA said regulators should look at the cost of providing the same function using another service, technology or band.

Professor Martin Cave, academic and consultant, has suggested a step-by-step methodology that can be used to incorporate this calculation into policy making[†] as shown in Figure 6.1.

[*] Smith System Engineering and National Economic Research Associates Economic Consultants (NERA). (1996). *Study into the Use of Spectrum Pricing* (Chapter 3), London, UK. http://www.ofcom.org.uk/static/archive/ra/topics/spectrum-price/documents/smith/smith1.htm

[†] For a detailed discussion of the theory of AIP and the practicalities of calculating prices, see Cave, M., Doyle, C., and Webb, W. (2007). *Essentials of Modern Spectrum Management* (Chapters 11 and 12), Cambridge University Press: Cambridge, UK.

There are other ways of calculating incentive prices. For example, one could find out how much revenue a firm gains from the use of each megahertz (MHz) of spectrum. However, this approach does not take into account the extent to which a particular application is dependent on one band in particular, and in any case one risks calculating the average value of a band for a user, rather than its marginal value.

As in auctions, it is generally considered bad practise to treat AIP primarily as a way of raising revenue for the government.

The success or failure of spectrum liberalisation in the public sector can be judged by whether formerly public sector spectrum has been assigned to the private sector. The problem in assessing this is that it has been only applied in a limited number of cases. Its most thorough implementation has been in the United Kingdom, but other countries have taken steps to impose spectrum pricing in some limited circumstances.

Countries That Have Implemented AIP

Only a handful of countries have based even some of their incentive spectrum prices on opportunity cost. These all have cultural ties to the United Kingdom.

Australia

Australia was one of the first administrations to introduce spectrum pricing, in 1995. The regime, known as an 'apparatus license system', is based on market prices but not on opportunity cost. The calculation is based on a formula whose inputs are frequency, geographic location, bandwidth and power.

Many other countries have implemented some forms of spectrum pricing, but without taking the further step of basing the prices on opportunity cost. Experts have argued that the reason many administrations do not wish to use AIP is that the complex econometric models are beyond the resources of many regulators. A regulator could hire consultants to do the work, but this is considered to be a major expense, especially if the process is repeated regularly. For example, despite otherwise being a leader in spectrum liberalisation, the New Zealand public sector only pay a small administrative fee for its use of spectrum in order to cover the office costs associated with the licensing process. Some private sector spectrum users pay an opportunity-cost-based fee, but this is based on auction revenues from similar bands.

In 2010, the Australian Communications and Media Authority (ACMA) decided to develop a methodology for introducing opportunity cost pricing into the 400 MHz band, which is commonly used for private mobile radios. It began by implementing opportunity-cost-based fees in three high-density areas of Australia: Sydney, Melbourne and Brisbane. The annual licence 'tax' in these areas is planned to eventually reach $199/kHz, and the first of five incremental increases to reach this level was implemented in 2012. No further increments have been levied since then.

In the second half of 2014, during a review of ACMA's 5-year spectrum plan, the regulator considered a second incremental rise in fees in the 400 MHz band, and whether to introduce opportunity cost fees to remote areas of Australia. In these areas, the opportunity cost is considered to be $0, as there is little demand in these sparsely populated areas. This would effectively be a tax cut as the incentive price for rural areas is currently set at A$4/kHz, although there is a nominal minimum tax of A$37.48 per user.

In 2012, the South African regulator, ICASA, introduced AIP for some fixed links and satellite services. Bahrain has also partially implemented AIP.

United Kingdom

The only administration to thoroughly implement AIP is the United Kingdom.

Spectrum pricing for both the public and private sector was first proposed by the UK Radiocommunications Agency (RA) in a consultation document issued in 1994, and was then advocated by the United Kingdom's conservative government in 1996 in a White Paper called 'Spectrum Management: Into the 21st Century'. Defending the White Paper in Parliament, President of the Board of Trade Ian Lang told the House of Commons that 'the Radiocommunications Agency [Ofcom's predecessor] has fully explored existing regulatory measures to manage the spectrum effectively. These are no longer sufficient so we intend to introduce spectrum pricing to augment them'.

One year later, New Labour won a landslide victory in a UK general election. Although spectrum pricing was not in Labour's 1997 election manifesto, it was retained in the 1997 Wireless Telegraphy Bill. In the bill, spectrum pricing was designed to be phased in, starting with the most congested frequencies. According to a House of Commons research document, 'safeguards will be introduced to protect small businesses, charities and essential services'. The RA's pricing framework was amended in 1998. At this point, spectrum prices were generally set at 50 per cent of a given band's full opportunity cost.

Five years later, Professor Martin Cave submitted his *Review of Radio Spectrum Management*. He argued that although the fundamental principle was sound, the price levels were too low to provide proper incentives. By 2005, Cave said in his *Spectrum Audit* of public sector spectrum that AIP so far had 'not led to spectrum usage being considered rigorously' by the public sector. To address this, Cave recommended that AIP should be applied across as many bands as possible, including radar bands and bands that are harmonised across NATO administrations.

His hope was that in the long run, the price paid for spectrum would reflect a 'spectrum value curve' based on the physical characteristics of a given band, rather than on how much the incumbent user can pay.

From its inception in the mid-1990s, spectrum pricing has also been used on private sector licences in the United Kingdom as a means of determining how much a licence holder should pay the regulator to protect it from interference. The idea is that charging the licence holder according to the market value of the spectrum is fairer than charging them according to the regulator's costs. All private sector spectrum holders in the United Kingdom pay AIP, aside from for bands that were purchased at auction. The principle was extended in 2013 when Ofcom proposed to apply full opportunity-cost-based fees to the GSM bands (900 and 1800 MHz bands) that mobile operators were given through an administrative process in the 1980s. These bands had subsequently been liberalised to allow more modern mobile technologies to use the band, and Ofcom wanted the fees to reflect the higher value that could be generated from the spectrum. In effect, Ofcom was proposing to more than quadruple the annual fees for the mobile operators. Predictably, the mobile operators challenged this calculation of annual licence fees and at the time of writing the issue has still not been resolved.

In 2010, Ofcom published a review into spectrum pricing.[*] It said that while it had not led to a significant re-farming of spectrum, it had 'met its primary objective of incentivising decisions that were more likely to lead to optimal use of the spectrum'. It also identified cases in which AIP has plausibly promoted more efficient use of spectrum and has not led to any adverse consequences. It emphasised that it is too early to properly evaluate the consequences of AIP.

It added that it maintains its view that AIP is an important spectrum management tool and that it 'expects to continue to apply it where appropriate for the foreseeable future'. Nevertheless, the term 'where appropriate' belies Ofcom's statement that each band should be managed according to its own circumstance, and that AIP cannot take into account the social benefits that can be generated from spectrum. On the role of AIP in the spectrum management toolbox, the document concluded that 'AIP is a valuable *complement* to spectrum auctions, trading and liberalisation and can usefully reinforce incentives from trading' [our emphasis].

Following are three case studies on the impact of AIP in the United Kingdom.

Military. According to the most recent data, the Ministry of Defence (MoD) pays Ofcom an annual fee of £155 million,[†] which is 58 per cent of all of Ofcom's income from spectrum fees.

Cave's 2002 *Review of Radio Spectrum Management* recommended that the UK MoD should pay full opportunity cost for its fixed links and mobile spectrum, but its charge for radar use would be increased

---

[*] Ofcom. (2010). *SRSP: The Revised Framework for Spectrum Pricing.* London, United Kingdom: Ofcom (http://stakeholders.ofcom.org.uk/consultations/srsp/summary).

[†] See addendum to document above: Ofcom. (2013). *Addendum to Consultation: Spectrum Pricing: A Framework for Setting Cost Based Fees* (http://stakeholders.ofcom.org.uk/binaries/consultations/cbfframework/summary/CBF_Addendum_to_Consultation_Nov13.pdf).

in line with maritime and aeronautical spectrum. In effect, this more than quadrupled the MoD's annual spectrum fee from £23 million to over £100 million.

Three years later, in his *Spectrum Audit*, Cave proposed further measures to slim down the MoD's spectrum assets. These included the extension of AIP and the introduction of Recognised Spectrum Access (RSA).

RSA is a regulatory fix that gets around the legal problem of Ofcom having no power to regulate Crown Bodies such as the MoD. In effect, RSA acts as a proto-licence where a regulator will protect a user's spectrum from harmful interference – and will levy charges – even though the regulator does not have the power to regulate the spectrum. RSA complements AIP as a means of enabling the military to sell its spectrum because it establishes the existence and ultimately the value of a given band.

It also instructed the MoD to come up with its own implementation plan for Cave's proposals.

In 2008, the MoD published its plan. It said that it would release and share spectrum, at least regionally, in the 406.1–430 MHz band in 2009. Its other priority was to look at ways to share the 3400–3600 MHz band. The MoD also announced its willingness to introduce RSA.

In addition to the 23 military bands that Cave had identified in his 2005 audit, the MoD said it was considering a review of its remaining 186 bands. The MoD was praised for its 'cutting-edge' proposals to share military spectrum.

Despite being cutting edge, these plans became beset by delays and 'pauses'. Nevertheless, the spectrum release plans were revived in the 2010 *Strategic Defence and Security Review*, and the MoD promised to publish its further plans in Spring 2011. The review had sought to generate 'in excess of £500 million' from selling spectrum and land.[*]

However, the 406.1–430 MHz band was absent from the MoD's relaunched plans. Instead, it focussed on the 100 MHz that could be

---

[*] HM Government, Cabinet Office. (2010). *Securing Britain in an Age of Uncertainty: The Strategic Defence and Security Review*. Norwich, United Kingdom: The Stationery Office.

extracted from the 3.4–3.6 GHz band. The MoD conceded that the eventual auction could take place as late as the end of 2012.

By the beginning of 2012, the auction date had been pushed back to 2015. In the meantime, Ofcom offered sharing opportunities to the private sector in the 3.4–3.6 GHz band. These were to be administered through Ofcom and would effectively incorporate short-term (3 year), regional leases. It said it could imagine sharing other bands in the future, such as the 870–872 MHz band, the 915–917 MHz band, the 1427–1452 MHz band and the 4800–4900 MHz band. None of these sharing schemes were ever taken up by the private sector.

The MoD also refined its plans to say it would hold auctions of the 3.4–3.6 GHz and 2.3 GHz bands by 2016 and 2014, respectively. The latter had not been mentioned by the MoD or Cave's reports for release or sharing prior to this announcement.

By mid-2013, the MoD had given up auctioning the frequencies themselves and passed Ofcom the responsibility to hold a combined auction of the 2.3 and 3.4 GHz bands on an exclusive basis. At the time of writing, Ofcom says that assigning these bands is a priority and that it expects to have done so by the end of 2015, although it may slip to the beginning of 2016.

AIP may well eventually prove to be an important part of spectrum releases to the private sector. Indeed, plans to share 870–872 MHz and 915–917 MHz are far advanced. Nevertheless, after 10 years of AIP in which over a billion pounds of public money has been paid for the MoD's access to spectrum, the only concrete output is the issuance of consultations and the employment of consultants.

Maritime sector. In the United Kingdom, the Maritime and Coastguard Agency (MCA), which is an executive agency of the Department for Transport (DfT), manages the use of spectrum for the maritime sector. In effect, it delegates most of these responsibilities to competent harbour authorities, although it does operate a Channel Navigation Information Service for the Dover Straights, which is one of the busiest sea lanes in the world.

Some frequencies used by the shipping industry are allocated by the International Telecommunication Union (ITU), and other frequencies are assigned by Ofcom. The MCA, or Ofcom's ability to regulate the use of spectrum by ships, is limited by the fact that the majority of the ships visiting the United Kingdom are registered in other jurisdictions.

Cave's 2005 audit recommended several measures to liberalise these highly regulated bands. For example, he said that Ofcom should review international allocations in the 156–158.5 MHz bands and 160.6–163.1 MHz bands. It should also, he said, implement AIP for navigational aid radar, coastal station radio and Differential Global Positioning System.

Cave's spectrum review, 3 years earlier, had given transportation uses a get-out-of-gaol-free card when it came to AIP. He said that although in principle all public sector users of spectrum must pay a fee based on the opportunity cost of a given band, in some bands the opportunity cost will be zero. This could occur in a situation where a band is globally harmonised for one particular service, and its use for another service would be impossible, such as for maritime.

Ofcom accepted the 2005 audit and spent 3 years preparing its proposals to implement AIP for the maritime and aeronautical sectors, which together occupy around 7 per cent of spectrum below 15 GHz. In July 2008, after having engaged some consultants to plan how AIP could be applied for the maritime and aeronautical sectors, Ofcom said that it would implement AIP in two phases. A first phase would see VHF communications paying AIP, using a similar methodology to that which it had established for business radios. A second phase would see the invention of a new system to implement AIP for radar and aeronautical navigation aids. It said that it had no intention of applying AIP to ship radio licences, aircraft or maritime or aeronautical distress channels. According to Ofcom's proposals, the MCA, or DfT, could pay AIP because these bodies have responsibility for the sector's use of spectrum, even though these bodies do not directly use the spectrum.

These plans changed in October that year. Two members of Parliament drafted an Early Day Motion in the UK's House of Commons condemning the proposals' impact on the Royal National Lifeboat Institution (RNLI), a charity which addresses the UK's maritime search and rescue needs. They argued that increasing the annual charge it pays Ofcom from £40,000 to £260,000 would 'threaten both local and national sea and mountain search and rescue charities across the country'.

Three days later, Ofcom amended its proposals to vastly reduce the fees the RNLI would have to pay.

Ofcom maintained that it may charge the DfT for the use of the spectrum. The shipping industry argued that any imposition of spectrum charges for the shipping would ultimately trickle down to the shipping industry. According to the Chamber of Shipping Deputy Director-General Edmund Brookes, this would 'be a disincentive to ships to use British ports' as no other administration does this.

The next summer, Ofcom decided to confine the application of AIP to the eight international channels used by port operators as demand exceeds their supply, and to the channels that are assigned by Ofcom but not internationally protected.

That December, Ofcom abandoned its proposals to charge AIP widely across both the maritime and aeronautical sector. Although maritime VHF channels and aeronautical navigation aids have paid AIP since 2010, both of these sectors now pay much less for the spectrum than it costs Ofcom to keep the spectrum cleared.

Aeronautical spectrum.   Communications in the global aviation industry are based on analogue technology, and its spectrum needs are met by the ICAO on an administrative basis. The majority of these spectrum needs come from Air Traffic Control services and in the United Kingdom, these are handled by various bodies depending on a given aircraft's purpose, flight path and current position. Ultimately, the Civil Aviation Authority governs aviation spectrum in the United Kingdom and participates on Ofcom's behalf at the ICAO and ITU levels.

In 2002, Cave recommended that a limited pricing regime should be phased in for aeronautical radars 'around the middle of the decade'. He anticipated that National Air Traffic Services and major airport operators should pay this fee. On-board navigation and communications systems were exempt from fees on the grounds that their opportunity cost is almost zero. Nevertheless, he did recommend that the Civil Aviation Authority should consider setting differential fees if UK-based users have a choice of different equipment.

Cave's 2005 audit also advocated the extension of AIP to civil aviation, on the basis that this would maximise the benefits of spectrum to aviation, and 'recognise and enable' other potential users of the bands.

Nevertheless, as there are few other users of these bands, the opportunity cost is likely to be low, so Cave recommended a conservative application of AIP to these bands. In order to deal with the panoply of organisations this would affect, Cave recommended the establishment of a 'coordinating body' that would note that 'algorithms which reflect impact on other spectrum users should be employed where this is not feasible'. Nevertheless, other users would only have to pay cost-recovery costs to Ofcom, and would possibly have to pay a penalty for unwanted emissions that radar systems cause.

He also argued that the Civil Aviation Authority should release some of the spectrum it used for radar systems in UHF television channel 36 (590–598 MHz).

Ofcom accepted his recommendations and started working on different ways to introduce AIP to aeronautical spectrum. These initiatives were strongly opposed by the International Air Transport Association, which represents civilian airlines. In 2008, a representative said that 'aviation demands integrated and consistent policy and regulation. It must also properly consider the benefits and the costs of the proposed regulatory changes. AIP does not meet these criteria'. The spokesperson pointed out that AIP would incur costs for the aviation industry without any benefits because changing their spectrum use has to be done at the ITU level. The spokesperson also questioned whether AIP would be permitted under the Single European Sky initiative, which ensures that the management and regulation of airspace are coordinated across the European Union.

Nevertheless, Ofcom persisted and in 2010 issued a consultation on the matter. The consultation attracted a record 227 submissions. Almost all of the responses opposed AIP. Many stakeholders portrayed AIP as a tax because it increased charges for airlines after they had already negotiated long-standing national arrangements which they could not terminate. The amateur flying community had particular venom for Ofcom's proposals. 'Daylight robbery for doing nothing', and 'a cynical abuse of government' were just some of the terms used in the public consultation.

These plans were abandoned along with similar plans for the maritime sector, and the aviation sector now pays cost-based and 'bespoke' fees.

It is interesting to note that improvements are being made in the spectrum efficiency of wireless equipment in the aviation sector. However, these are taking place through initiatives from ICAO and Eurocontrol. The latter is an organisation which aims to integrate Europe's fragmented 'Functional Airspace Blocks'.

Conclusion

AIP – and liberalisation in general – focus on maximising the amount of spectrum that is available to the private sector by providing incentives for the public sector to vacate it.

However, more recent initiatives have focussed on finding a way to share spectrum between all stakeholders. These new approaches are explored in more detail in the final section of this book. Nevertheless, it is worth briefly noting that the influential documents that have proposed new sharing approaches have also retained spectrum pricing in their list of tools that regulators can use.

The European Approach

In 2009, the intergovernmental group that advises the European Commission on spectrum matters, the Radio Spectrum Policy Group (RSPG) issued its final opinion on public sector spectrum. It stated that the public sector holds 'significant amounts of spectrum' and that it should use it as efficiently as possible so that the spectrum can be used for additional public services, or for non-public services.

Echoing Cave's views on the value of the market in spectrum, the opinion advocated the use of trading and pricing to ensure that the 'right amount of spectrum' was used by the public sector. It argued that fees would alert public officials to the opportunity cost that comes from using spectrally inefficient technology. Its endorsement of pricing and market mechanisms to incentivise the efficient use of public sector spectrum did not encompass the maritime or aviation sectors. It explained that this was because these services required global harmonisation.

The opinion established three principles for public sector use of spectrum. These were technology and service neutrality, that there should no presumption that the public sector is a more important user of spectrum than the private sector, and that spectrum sharing between the public and private sector should be considered.

In order to facilitate more sharing, the opinion recommended a series of measures that member states should consider implementing. These included thorough audits of spectrum use and the granting of different categories of user rights. These measures should be introduced gradually 'on a case-by-case' basis, taking into account harmonisation, interference and 'macro-economic' aspects, the opinion said.

The US View

One year later, the US National Broadband Plan also advocated the parallel application of spectrum pricing and clearance for the public sector, and also spectrum sharing between both the public and private sectors.

The plan recommended that 'Congress should consider granting authority to the FCC to impose spectrum fees on licence holders and to NTIA to impose spectrum fees on users of government spectrum'. These fees, it said, should be based on a given band's opportunity cost in order to 'mimic the functions of a market'. It added that fees should only be imposed on spectrum whose licences are not service neutral, as these are the bands that are insulated from the market.

Subsequently, these ideas have been developed by US lobbyists and academic institutions. A recent idea by the Technology Policy Institute posits that all public sector spectrum should be owned by a 'Government Spectrum Ownership Corporation', and this corporation should be able to lease spectrum to public sector users at market rates.

Opportunity-cost-based incentive pricing is a part of the National Broadband Plan's strategy to release another 500 MHz of spectrum for mobile broadband by 2020. The majority of that spectrum will be found through re-farming from other private sector uses of spectrum (such as broadcasters), and by sharing arrangements with the Department of Defence.

Both the RSPG opinion and the proposals in the US National Broadband plan show the theoretical strength of the AIP argument, and how this has convinced international policymakers at the highest level. This is particularly true in the United Kingdom, but the policy faced significant obstacles and has enjoyed only limited success. These practical difficulties perhaps explain why the theoretical enthusiasm for AIP has resulted in actual implementation in only a handful of countries.

7

Broadcasting*

When Coase wrote in 1959 about applying market forces to the use of spectrum, he was talking mainly about television. TV was then the 'biggest beast in the jungle' – the highest value use of the airwaves. However, Coase's vision has been applied most successfully to the mobile industry, which became the 'biggest beast' in the 1990s.

Broadcasting continues to occupy large amounts of valuable spectrum below 1 GHz that has been a key target for liberalisers trying to improve the efficient use of the airwaves. But serious attempts by national policymakers to liberalise broadcasting spectrum have been notable by their absence. Only two countries have even attempted this: the UK's efforts are on hold or moribund and the incentive auction in the United States was due to be completed the year after this book was published.

In this chapter, we explain why it has been so difficult to apply market mechanisms to broadcasting spectrum and why most countries have not even tried. We will be looking particularly at the social policy goals associated with broadcasting, the complexities of the TV market and its wide national variations, as well as the large national investments in the digital terrestrial television (DTT) platform.

We will also examine why the UK's attempt to liberalise spectrum policy failed, and consider the long-term trends in consumer and viewer behaviour which could see terrestrial television diminish in importance, perhaps creating more leeway for a liberalised approach.

Liberalisation and the Development of the TV Market

The first section of this chapter looks at the development of the TV market – a story of growing competition not only between broadcasters but also between TV platforms. We consider how this affects the

---

* This chapter was written with help from Catherine Viola.

arguments for and against the liberalisation of the spectrum used by the broadcasting sector.

'Technical Restrictions': An Oversimplification

In Chapter 2, we discussed Coase's interpretation of why, in the 1920s, spectrum was not subject to market mechanisms like most other scarce resources. He argued that policymakers were confused by the ephemeral nature of the airwaves and uncertain about their cultural importance.

There was another influential thread running through policy making, particularly in the United Kingdom. It was widely believed that 'technical restrictions' meant that broadcasting was a natural monopoly. The accepted view was that you could have only one radio station in each country, if they were to be of acceptable technical quality and cover the whole country. This perspective was strengthened by observing the 'era of broadcasting chaos' in the United States in 1926.

A natural monopoly would be abused by a commercial company: the best response was public ownership and so the concept of Public Service Broadcasting (PSB*) was born. It is most popular in Europe, but also developed in Asian countries such as India and Japan, in New Zealand and Australia, and in South American countries such as Brazil, Argentina and Chile. In some, including the United Kingdom, the state control of a media institution was frowned upon, so the BBC was run as an independent body, at arm's length from government.

The perceived technical restrictions were often referred to as a 'happy coincidence'. They gave substantial public control over the development of the new broadcasting medium, and this has had wide support from the public and across the political spectrum – the BBC in particular continues to be seen as a world leader in broadcasting.

---

* In this book, we define PSBs as being broadcasters *owned* by governments, be that a national or regional assembly. Although owned by a government, a PSB may not necessarily be funded by that government and is not usually controlled in any direct way by that government. We accept that many privately owned broadcasters also have public service commitments, but in this book we refer to these as commercial broadcasters.

In fact, it is doubtful that technical restrictions ever forced the governments to have such a large role in broadcasting. Radio was not operated as a monopoly in the United States, which was the pioneer in the early days of the medium. In Europe, non-state broadcasters were widely listened to in the interwar period. The best known was Radio Luxembourg which started services to England in 1933 but also broadcast in French and German. Its English services used first one frequency which was not agreed by the relevant international body, then switched to another which was similarly unauthorised. This provoked intense protests from the British Post Office. Radio Luxembourg continued broadcasting and developed an audience of four million by 1938 before closing down during wartime.* Prior to 1933, there were other commercial stations broadcasting into the United Kingdom from abroad, such as Radio Normandie and Radio Lyons, making the natural monopoly claims highly questionable from the start.

Depending on the country and how far back you go, there was probably a grain of truth in the claim that technical restrictions severely limited the number of possible TV or radio channels, but it is not the full picture (indeed such arguments persist in many countries). A more rounded explanation is that it was a convenient and simple way of justifying a complex argument. Governments around the world were nervous about the potential power of radio and television and wanted to have some control over its development. The use of radio in Nazi propaganda reinforced those fears.

Broadcasting was also enormously important to governments as an addition to the national infrastructure: it is the most effective means of communicating with the whole population in the event of a national emergency. Broadcasting is a key repository of power for the modern state: in a *coup d'etat* the plotters go for the president first and the TV station second. Some European countries subsidise coal production because they want to keep control of at least one source of energy production; similarly, some governments want to keep control of at least one type of communications infrastructure, typically terrestrial broadcasting.

---

* Information provided by the RTL group: On Air. *History of Radio Luxembourg and Its English Service* (http://www.radioluxembourg.co.uk/?page_id=2).

An even more complex argument is in the realms of arts and cultural policy. The PSBs produce content specific to national markets and cultures, as well as services which cover the whole country. These boost national, geographic, social and ethnic cohesion. The production of programmes, particularly for TV, is an important market in economic terms in some countries. It is a source of employment which brings in tax revenues and boosts exports.

The technical restrictions justification was the simplest way of winning the argument on broadcasting policy, but it tended to hide these other benefits which became a more public part of the debate from the 1980s onwards.

The Rise of Platform Competition

The natural monopoly justification got weaker and weaker as the years went by. In 1955 Europe's first commercial TV network[*] was launched in the United Kingdom, and over the succeeding decades a 'dual system' combining PSB with private television stations was adopted across much of Europe.[†] In terms of spectrum policy, this was a confirmation that the part of the airwaves used by TV was not a natural monopoly and could be shared with commercial companies.

Soon there was platform competition as well as programming competition. Cable TV and satellite TV took off in the 1980s, with about 57 per cent of US TV households using cable by 1990[‡] with satellite lagging behind, securing 4 per cent of viewers by 1996.[§] Cable and satellite TV grew during the same period in Europe but overall the pace was slower. By 2002, the number of cable TV households was

---

[*] Donders, K., Pauwels, C., and Loisen, J. (2013). *Private Television in Western Europe: Content, Markets, Policies*. London, United Kingdom: Palgrave Macmillan, p. 71.

[†] For example, France adopted it in 1982 and Germany in 1984. Commercial TV came to Eastern Europe during the 1990s after the fall of the Berlin Wall.

[‡] Figures from US National Cable & Telecommunications Association.

[§] Cable and Other Pay Television Services. *Reference for Business: Encyclopedia of Business*, 2nd ed (http://www.referenceforbusiness.com/industries/Transportation-Communications-Utilities/Cable-Other-Pay-Television-Services.html).

estimated at 27 per cent, with satellites at 15 per cent while terrestrial TV households accounted for 58 per cent of the European total.[*]

This meant that consumers could now choose to have their TV delivered using the traditional terrestrial means, typically using the broadcasting section of the VHF and UHF bands (174–230 and 470–862 MHz), or by satellite using Ku band (12–18 GHz) for smaller dishes or C-band (4–8 GHz) for larger dishes. By the second decade of the new millennium, there was healthy competition in TV platforms.[†] In 2009 research indicated that 55 per cent of Europeans used pay-TV platforms, overwhelmingly cable and satellite; in the United States, it was 87 per cent and worldwide it was 49 per cent.[‡] Since 2009, there has been further growth in another platform: TV delivered via the Internet (IPTV), exemplified by video streaming services such as Netflix.

Spectrum Efficiency

Terrestrial television occupies a prime piece of spectrum where the propagation characteristics mean that large amounts of information can easily be carried over reasonable distances. These are frequencies which are much in demand for other uses, principally wireless broadband.

There is a question mark over the efficiency of DTT when compared to other technologies such as cable and satellite. In the United Kingdom in 2015, for example, DTT services using the latest DVB-T2 technology offer no more than 10 high-definition (HD) channels in the spectrum available to them, whereas competitor satellite services could offer over 70 HD channels and cable services have over 40.

There is a further pressure on DTT. Research shows that teenagers are less keen on live TV – the mainstay of the DTT platform – preferring to watch TV via catch-up services on the Internet.

---

[*] www.parliament.uk. Select Committee on Welsh Affairs: Appendices to the Minutes of Evidence. Annex: Digital TV (http://www.publications.parliament.uk/pa/cm200203/cmselect/cmwelaf/95/95ap03.htm).

[†] In many countries, this is a regulated competition because cable companies and sometimes satellite platforms are required to carry terrestrial programming: the 'must carry' obligation.

[‡] IDATE. *TV Facts and Figures* (2010).

Today, most viewing is 'linear', that is, occurring at the time of the live broadcast. Non-linear viewing, including using popular Internet catch-up TV services such as the BBC's iPlayer in the United Kingdom, still accounts for less than 10 per cent of all viewing. But this could change, as smart TVs that allow broadcast content to be viewed on demand via the Internet increase in penetration and as broadband speeds rise to improve the experience of using such services. If non-linear viewing were to become the norm, then a shift away from DTT broadcasting might be justified, but for now the future direction is still unclear.

How consumers choose to view broadcast content is also evolving. Personal communication devices are proliferating, offering consumers several choices for how they watch broadcast (and catch-up) content. Teenagers and young adults are increasingly using laptops, tablets or mobile phones to watch TV, rather than traditional large-screen sets. This could eventually push broadcasting delivery towards broadband or wireless IP (LTE) platforms, but again it is too early to determine if, when, or to what extent this might happen.

One would imagine that once spectrum liberalisation was implemented, DTT would struggle to survive or end up as a third-best option at the cheaper end of the market. The platform's decline would be because it has less spectrum (and thus capacity) available than satellite; the channel offering is inferior to satellite and cable; and the younger audience seem to be moving onto the Internet. However, this is far from the truth. DTT looks likely to survive for decades to come, and in the next part of this chapter we examine why.

Why Is DTT Thriving?

Varying penetration of DTT.   The first issue is that the use of DTT varies widely from country to country.

In some countries, there is a very high take-up of cable and/or satellite TV subscriptions, as Figure 7.1 shows, and only a small proportion of the population relies on the terrestrial broadcasting platform. Within Europe, cable TV is very well diffused in Germany, the Benelux countries, Sweden and Eastern European countries such as Bulgaria, Poland and Romania. North America and some parts of Asia also have a high penetration of cable TV in urban areas.

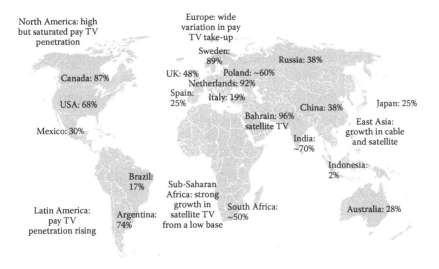

North America: high but saturated pay TV penetration

Canada: 87%

USA: 68%

Mexico: 30%

Europe: wide variation in pay TV take-up

Sweden: 89%

UK: 48% Poland: ~60%
Netherlands: 92%
Spain: 25% Italy: 19%

Russia: 38%

Bahrain: 96% satellite TV

China: 38% Japan: 25%

East Asia: growth in cable and satellite

India: ~70%

Indonesia: 2%

Brazil: 17%

Sub-Saharan Africa: strong growth in satellite TV from a low base

South Africa: ~50%

Australia: 28%

Latin America: pay TV penetration rising

Argentina: 74%

Figure 7.1   Global trends in pay-TV penetration (DTT delivers paid-for content in some countries). (From *PolicyTracker*, based on Worldscreen figures. With permission.)

Satellite is well established in mature TV markets and is growing strongly in emerging regions, particularly in large sparsely populated areas where cable is uneconomical to roll out. The take-up of cable and satellite has been driven by the adoption of the paid-for content provided on these two platforms, hence Figure 7.1 refers to the penetration of pay TV. Paid content is also delivered by DTT, but cable and satellite are by far the more popular platforms for these services.

Within just a few years, terrestrial TV is expected to be the primary access mode for only around a quarter of households. Nonetheless, analogue and DTT is still dominant in a range of developed and emerging markets. In Spain and the United Kingdom, more consumers use DTT than any other platform (especially on second sets, i.e. in bedrooms). Free-to-air digital broadcasting in the United Kingdom is very popular with consumers and can be accessed without subscription via both DTT (Freeview) and satellite (Freesat) platforms. Research commissioned by Ofcom and Freeview has shown that UK consumers place a high value on being able to receive the wide range of free-to-air content that is available via Freeview.[*]

---

[*] Ofcom's 2008 PSB review found that over 75 per cent agreed that 'it is important that TV is available to everyone', while in Freeview's 2011 consumer survey 83 per cent agreed it was important to have access to digital TV without subscription.

In some emerging markets, there are still monolithic, state-controlled TV broadcasters, sometimes coupled with an environment of strong restrictions on broadcasting. In extreme cases such as Sudan there are no privately owned TV stations, and broadcasting reinforces government policy as well as being subject to military censorship.

This snapshot of pay-TV penetration suggests that, from a purely market-driven perspective, DTT could have a case to answer regarding its continuing use of sub-1 GHz spectrum, at least in certain markets. The rising take-up of pay TV in emerging markets suggests that this will increasingly become the case.

However, the picture of pay-TV adoption is by no means uniform, either between or within regions. In the mature European TV market, there are large country variations, and there are also big differences within and between emerging regions. These very different starting points significantly complicate any debate into the longer term use of the TV bands. Any decisions would require international coordination, and this means that countries with little DTT usage are held back by neighbours where the platform predominates.

Universal service.   The social cohesion function of PSB only works if everyone can receive it, and in many countries including the United Kingdom, Spain and Italy, terrestrial television has been the way of meeting this universal service obligation.

When television began in the 1950s, networks of high-power transmitters that take advantage of the excellent propagation characteristics of VHF/UHF spectrum were generally the most effective way to deliver near-universal TV coverage to citizens. Consequently, many countries have very large installed bases of rooftop aerials and TV sets capable of receiving terrestrial TV signals, and with this equipment installed households can access the PSB channels for free.

It is difficult for alternative broadcast technology platforms to match terrestrial TV's universal coverage and free access proposition for PSB, and this helps to entrench DTT's hold over the UHF airwaves. Cable is not suitable for universal access, as it is uneconomical to roll out in rural and remote areas. Satellites have massive footprints and so can provide excellent coverage, including of rural and remote communities. Satellite systems also have plenty of capacity to support the future expansion of HD programming. But fewer than one-third of households globally already have satellite dishes and receivers installed, and a big (but not insurmountable) investment would be needed to switch the remaining households from DTT.

Eventually, IPTV could provide another option for PSB services, although this is unlikely to be viable for at least 10 to 15 years in many countries. Today IPTV has several limitations, including the high costs of delivering high-capacity video and insufficient bandwidth for delivering HD services. These difficulties will be surmounted over time, for example as multicast technology is adopted and high-speed broadband infrastructure, such as fibre to the cabinet or fibre to the home, becomes more prevalent. However, as a candidate for PSB delivery, IPTV will only become attractive once suitable broadband infrastructure is widespread and there is a high household penetration of broadband services. If these conditions are met, installing a suitable IPTV receiver would be the only additional cost, but for many countries, particularly those in the developing world, near-universal broadband is a very distant prospect (Figure 7.2).

Achieving universal coverage for PSB is arguably not the most substantial barrier to migrating DTT-only households to alternative platforms. Satellite would be a suitable alternative in that regard. More problematic are the overall costs associated with substitution, which are driven both by the proportion of DTT-only households and by the substitution costs per household. For satellite TV, professional installation of the dish is preferred in many countries, adding an extra cost beyond the price of the equipment, making it more expensive per household than other options. That being said, in a country yet to make the transition to digital television, the cost differential between replacing existing equipment with a terrestrial set-top-box as opposed to a satellite dish and satellite set-top-box is minimal.

But the penetration of other platforms could eventually reach a tipping point at which the economic benefits of migrating the remaining DTT-only households to another platform such as satellite or IPTV would outweigh the costs. Forcing the issue before this

Technology	Coverage capability	Free at point of access?	Comment
DTT	Networks of high-power transmitters using low-frequency UHF spectrum can provide near-universal coverage	Yes. Requires compatible TV and rooftop aerial, but large installed base for legacy reasons	DTT often used as access platform for extra TVs in the home, even if it is not the access mode for the primary TV
Cable	Best suited to urban environments. Roll-out uneconomic for sparsely populated areas	No. Requires a subscription, although PSB channels provided at no extra cost	Accounts for a large proportion of primary TV access globally but is not suitable for universal access
Satellite	Extensive satellite footprints provide universal coverage. Effective for rural as well as urban areas	Yes, in some circumstances, but requires satellite dish and set-top box, plus installation costs	Satellite is already used for rural in-fill TV coverage in some countries with challenging terrain
IPTV (managed service)	Broadband infrastructure needs upgrading to multicast technology and higher speeds to be a complete substitute for other platforms. Suitable infrastructure and broadband penetration levels are too low in near term to provide a universal service	No. Requires a broadband subscription and an IPTV receiver	Could emerge as a substitute for DTT once superfast broadband is widely adopted

Figure 7.2   The suitability of the main broadcasting platforms for delivering PSB services.

natural threshold is reached would likely result in a net cost rather than benefit. For this reason, it would seem unlikely that countries with a low cable or satellite penetration would be willing to contemplate switching off DTT, even if their neighbours were pressing to do so. This could mean that the timing of any future change of use of the DTT bands is driven by the countries that are slowest to adopt subscription TV.

Citizens' entitlement to free access to PSB may also require future rethinking. Like DTT, satellite would allow consumers to receive PSB for free at the point of access, provided they had the right equipment. Cable and IPTV, on the other hand, generally require a household to have a subscription before they can receive the service. If the householder then terminated their subscription, they would typically no longer be able to access PSB.

In short, many countries have no cost-effective alternative to DTT which meets all the PSB requirements, unless they are lucky enough to be one of those countries such as the Netherlands, which has near-ubiquitous cable. Even in the Netherlands, with 92 per cent of pay-TV subscribers, the government welcomed DTT as a competitor platform to challenge the domination of cable. It had achieved nearly a million customers by 2009, attracted by competitive pricing and getting a reasonable range of programming for second sets, boats, caravans and holiday homes.* Even in this most cabled country, DTT is providing a social function.

DTT's product life cycle.   Governments and regulators have also been reluctant to interfere with policy on broadcast spectrum because digital transformation in terrestrial TV is still a relatively new phenomenon and their focus has been on completing analogue switchover. The DTT broadcasters have invested millions of dollars in upgrading their platforms to more spectrally efficient digital technologies, enabling substantial bandwidth to be released for mobile broadband services in the 'digital dividend' bands.

As a technology, DTT is at an early stage in its life cycle and the potential for further development discourages governments from adopting a liberalised policy which could favour satellite, cable or IPTV. The introduction of new transmission and audio-visual compression technologies, such as DVB-T2 and MPEG-4/H.264, is allowing the platform to offer high-definition services (HDTV) although the number of programmes it can support is far fewer than cable and satellite. A third generation of video

---

* *PolicyTracker.* (2010). The Surprising Success of DTT in One of the World's Most Cabled Countries (https://www.policytracker.com/headlines/the-surprising-success-of-dtt-in-one-of-the-worlds-most-cabled-countries).

compression technology, known as high-efficiency video coding, will become available and is being rolled out from 2014 onwards, and could double the number of HD channels supported on DVB-T2 multiplexes over the next decade and permit the introduction of 4K, ultra-high-definition (UHD) services. The DTT services are also being improved by the use of single-frequency networks (where all the transmitters use the same frequency and thus require far less spectrum).

To summarise, DTT still has a lot to offer as a competitor to cable and satellite. It is not a technology which is easy to jettison because it is 'on its way out': in some countries it is an extremely attractive prospect.

However, some market trends suggest a move away from DTT. While there is a clear road map towards more spectrally efficient broadcast technologies over the next 20 years, a significant shift in viewer demand towards HD, 3D and UHD TV would increase the amount of capacity required to deliver each DTT channel. Even with the benefit of new multiplex technologies, this is likely to drive up the need for bandwidth – or else face reducing the number of channels offered. At present, it is uncertain how demand for these new programming formats will evolve, and what overall impact this will have on DTT's appeal to consumers.

UK Attempts to Liberalise TV Spectrum

In the first wave of enthusiasm for spectrum liberalisation, the only country to make any attempt to apply market mechanisms to broadcast spectrum was the United Kingdom.* In 2006 the regulator Ofcom proposed pricing broadcast spectrum based on the opportunity cost principle (i.e. how much the frequencies would have been worth to another user). This is known as administered incentive pricing (AIP) and had already been applied to the military (see Chapter 6).

Exactly how much broadcasters would be charged was never explicitly stated but the Ofcom statement implied annual charges for TV spectrum of around £16 million for Channel 4 and £65 million

---

* The US incentive auction was a later development, being proposed from 2010 onwards. It is also relevant to this discussion but is covered in Chapter 23.

for the BBC,* although the BBC had earlier estimated around £300 million. The AIP proposals would also apply to BBC radio.

Ofcom argued that paying for spectrum makes licence holders aware of its market value, giving them an incentive to use a limited resource more efficiently:

> The BBC's radio arm, and the DTT and DAB multiplex operators, have hitherto had free spectrum [...] The opportunity costs of their marginal use of spectrum, or other options for delivering content, have not automatically figured in their decisions.

In other words, if you get free electricity, you do not switch the lights off.

At the time, the example often discussed was: 'What if the BBC wanted to set up a city TV service?' As it paid nothing for DTT spectrum, terrestrial TV would be the natural choice. But this hides the true cost: using DTT would prevent other services using the spectrum. Some of these services – such as mobile – might bring a higher value to the economy, so in economic terms giving the BBC free DTT spectrum paints a misleading picture. If the BBC delivered city TV via satellite, they would be charged by the satellite operator; if they delivered it via the Internet, they would have to pay for the servers and bandwidth.

> If DTT spectrum was priced, then the BBC has to make a decision about the most cost-effective way of reaching its target audience. If it was a small city it seems likely that Internet streaming may be a more cost-effective option. As stated, '[I]f only spectrum is discounted a broadcaster is likely to use more spectrum than would be efficient, and invest too little in other inputs'.†

But Ofcom's persuasive arguments have so far come to nothing. It had planned to bring in pricing for TV spectrum in 2014, but in 2013

---

* See Ofcom. (2007). Future Pricing of Spectrum Used for Terrestrial Broadcasting. London, United Kingdom: Ofcom, p. 21 (http://stakeholders. ofcom.org.uk/binaries/consultations/futurepricing/statement/statement.pdf). Also see *PolicyTracker*. (2007). UK Presses Ahead with Market Regime for Broadcast Spectrum (https://www.policytracker.com/headlines/uk-presses-ahead-with-market-regime-for-broadcast-spectrum).

† Ofcom. (2007). Future Pricing of Spectrum Used for Terrestrial Broadcasting. London, United Kingdom: Ofcom, p. 48.

it was forced to delay the proposal by 7 years, until 2020. Ofcom said this was a firm commitment, but a 7-year delay meant implementation was at least two parliaments away: in political terms that is the equivalent of saying it may never happen.

The Need for International Coordination

Ofcom's explanation highlights one of the obstacles to spectrum liberalisation – the slow-moving and interconnected nature of the market. The regulator said national DTT broadcasters were facing 'a unique set of circumstances', namely, the move out of 700 MHz and into 600 MHz. Freeing up 700 MHz for mobile would take extensive international coordination and a replanning of the DTT platform involving significant coordination between multiple stakeholders. This meant that 'AIP is unlikely to encourage more efficient use of spectrum in the short to medium term'.*

There is also a technical issue here: broadcasting and mobile are very different services, TV being high power and cellular being lower power. Unlike mobile, some TV networks also use interleaved spectrum (i.e. different frequency ranges in different parts of the country). Reallocating TV bands to a very different use therefore requires a great deal of management: one service could not be simply replaced by another similar service – the scenario which would be best suited to a market approach.

In other words, the only way for significant TV spectrum to move to higher value mobile uses was via a traditional command-and-control method. To create the economies of scale needed to produce mobile equipment for 700 MHz, this had to be harmonised on a regional or even global scale both to create sufficiently large markets and to prevent interference. By 2020, command and control will have set the wider framework, and Ofcom then believes there is scope for market mechanism to work in a more limited fashion. The AIP may then lead to the release of smaller amounts of DTT spectrum, but these are unlikely to be valuable for other uses because they are not part of harmonised bands.

---

* *PolicyTracker.* (2013). Ofcom's Broadcast Spectrum Proposals May Send the Wrong Signals (https://www.policytracker.com/headlines/ofcom2019s-broadcast-spectrum-proposals-may-send-the-wrong-signals).

Complexities of the TV Market

The intricate ecosystem of the TV market throws up other problems for liberalisation. The AIP seems to work better in public services such as the military. If an army budget is £800 million, the government sets a price for spectrum and gives the generals an extra £100 million to spend. If they use less spectrum, they keep some of the £100 million to spend on other things, such as tanks or guns.

Funding PSBs

TV is more complex. First, most PSBs operate in a highly constructed market designed to provide them with sufficient funding while ensuring they also have commercial competition. They may be funded by a share of advertising, where the number of possible competitors is limited by government policy, by direct grants from the state, by a licence fee paid by everyone with a television, by membership fees or by a combination of these.

Governments usually have a hand in setting these funding frameworks, either through the licence fee settlement (as in the United Kingdom) or through setting the level of grants or advertising percentage. Unlike the public sector, it is rarely a case of governments directly giving money to broadcasters. In the example above, if an army uses the same amount of spectrum as last year, they receive £100 million from the Department of Defence and give £100 million back to the spectrum regulator, which gives it to the government. There is no net loss or gain.

If the army uses less spectrum, they keep, say, £30 million for new guns. The spectrum regulator gets £30 million less but can sell the returned frequencies for at least £30 million if the AIP is reasonably accurate. This new use of new spectrum will also stimulate the economy and generate new tax revenues, so either way the state should benefit in the long term.

AIP for the military works so well because it is money going from one part of government to another: the government is paying itself for the spectrum and getting that money back. With broadcasting, the relationship is rarely so direct. In those countries with licence fees, this is often set by the government but paid by individual TV viewers, unlike military budgets which come from a share of overall tax revenues. Members of the public rarely notice accounting procedures – like AIP – used on a small part of the government's overall budget, but they would certainly notice if spectrum pricing put up their licence fee. Furthermore, asking TV viewers to pay for a change in government economic policy does not seem reasonable.

That problem could be resolved by governments making a direct payment to broadcasters to cover the new AIP charges. However, that was not what was proposed in the United Kingdom: the BBC called AIP the 'spectrum tax', arguing that it would lead to an increase in the licence fee and 'remove much-needed cash from the UK's creative sector'.* Ofcom was proposing setting a tax that might end up being paid by individual viewers, and it did not have the power to compel the government to give the BBC extra funds to meet the spectrum costs. Ofcom suggested that broadcasters could resolve this problem by cutting costs or raising revenues,[†] adding fuel to the claims that this would hit programming budgets.

Professor Martin Cave, who first mooted these proposals in his 2002 report,[‡] argued that charges should be phased in over several years to minimise any negative impact but this failed to convince the broadcasters.

Competition

A further complication is the nature of competition in the broadcasting market. First, PSBs are competing with commercial broadcasters. If both were to be charged equally for their use of spectrum, commercial broadcasters could put up their advertising rates, but this is not possible for ad-free PSBs who might therefore have to cut programming budgets.

---

* *PolicyTracker*. (2013). UK Broadcaster Urges Regulator to Waive "Spectrum Tax" (https://www.policytracker.com/headlines/bbc-calls-on-regulator-to-waive-201cspectrum-tax201d).

† 'PSBs may take steps on their own initiative to offset any material cost increases, either by reducing costs or increasing revenues associated with their broadcasting activities' (Ofcom, *Future pricing of spectrum used for terrestrial broadcasting*, 2007, p. 21).

‡ Cave, M. (2002). *Review of Radio Spectrum Management* DTI, London, UK.

Second, any broadcaster which depends on advertising is competing with other media. If AIP forces up the price of TV slots, advertisers may move to newspaper campaigns, direct mail or more likely the Internet, which has been eating into TV revenues for years, a very sensitive subject for politicians and the TV industry. So it is certainly possible that AIP could reduce advertising revenues and therefore programming budgets for both PSBs and commercial broadcasters.

If we compare this to AIP for public sector services, the competitive position is much simpler. The security of the nation does not depend on creating the right level of competition between state armies and their (hopefully non-existent) rivals in the private sector. Unlike broadcasters, armies do not compete for customers, so there is no market there which spectrum pricing might upset.

Overall, charging AIP for the UK's broadcasting spectrum came to be seen as difficult to execute and politically toxic. It may lead to a direct tax increase for individuals and it roused the ire of that most powerful of lobby groups – the broadcasters. No surprise that it ceased to be pursued with any vigour in the United Kingdom and failed to gain political traction elsewhere in Europe.

Professor William Webb was instrumental in developing Ofcom's spectrum policy during this period as their head of research and development. He said discussions on applying market forces to broadcasting took place as early as 2002 'but politics and the public service element of broadcasting always got in the way. Broadcasters argued that they were obliged to serve a large percentage of the population under law'.

'Politicians rarely saw the advantage in acting against this sector for little immediate gain', said Professor Webb, '[a]nd so, over a decade on, we still do not have any real market forces applied in this area with little change in sight'.*

The Long-Term Future of Terrestrial TV

In the early years of the new millennium, liberalisers hoped that market mechanisms could be used to strike a balance between TV spectrum demands and those of other services. A study commissioned

---

* Interview with the authors, December 2014.

by the European Commission (EC) in 2004 advocated using AIP to encourage analogue TV switchover: 'To promote spectrum efficiency broadcasters should face the marginal opportunity cost of their use of spectrum, just as for example they pay an unsubsidised price for electricity and other inputs they use'.* In fact digital switchover (DSO) was completed by administrative assignments in the European Union and in all other countries. As we have seen the UK's attempts to use AIP to help define the use of the UHF band have also stalled.

It is very likely that the long-term use of the UHF band will also be decided in an administrative manner. In recent years, the EC has not suggested market mechanisms as an answer and in 2014 it asked the former head of the World Trade Organization Pascal Lamy to try to find a road map, which was agreeable to the mobile industry, broadcasters and governments. He suggested a compromise known as the '2020–2030–2025' formula.† This calls for the 700 MHz band to be dedicated to wireless broadband across Europe by 2020, plus or minus 2 years. Terrestrial broadcasters should have regulatory security and stability in the remaining UHF spectrum below 700 MHz until 2030, and there should be a review by 2025 to gauge market and technology developments.

This compromise focusses on the idea that linear TV services will be the dominant mode of audiovisual consumption for the foreseeable future, while non-linear services, particularly those delivered over the Internet and broadband connections, will continue to grow from their relatively low level.

So the future use of the UHF band should be resolved by an administrative reassignment, not market mechanisms, according to Lamy, and this has been mirrored by announcements from other European countries. Finland, for example, recently has announced that all free TV

---

* Burns, J., Marks, P., LeBorgne, F., and Rudd, R. (2004). *Implications of Digital Switchover for Spectrum Management.* Prepared for the European Commission, Brussels, Belgium; (DG Information Society) by Aegis Systems, Independent Consulting, and IDATE. p. 12.

† Lamy, P. (2014). *Results of the Work of the High Level Group on the Future Use of the UHF Band (470–790 MHz).* Report to the European Commission   (http://ec.europa.eu/information_society/newsroom/image/pascallamysreportonthefutureuseoftheuhfband_8423.pdf).

channels would be accessible using current TV sets until 2026. In the United Kingdom, Ofcom is undertaking a review of the long-term strategy for the UHF bands, and proposes a continuation of DTT for at least the next 15–20 years.

It is clear that the continuing rise of cable, satellite and IPTV around the world will continue to fuel the debate over the ongoing use of the UHF airwaves by DTT and the platform's usage will become increasingly hard to justify. Calls from the mobile sector for access to more sub-1 GHz spectrum and their ability to deliver quasi-broadcast services using technologies such as LTE Broadcast are already beginning to act as catalyst for the industry to take a serious look at this issue, and as unused capacity in the digital dividend bands diminishes, these calls will only get louder.

But any change to the status quo will inevitably be slow. Earlier ideas about spectrum liberalisation are coming up against the realities of TV investment cycles. Broadcasters are making long-term investment commitments to their DTT platforms, and it is unreasonable for them to be given anything but long-term notice of any proposed change in their access to the UHF band. Many are already investing in a second generation of DTT technology, and it could take up to a decade for all consumers to universally replace their TVs or set-top boxes and so to benefit from these upgrades.

The exception may be in countries yet to make the switch to digital broadcasting. Should these countries delay the introduction of DTT until such time as cable, satellite, IPTV and LTE broadcast are able to deliver access to 90 per cent plus of viewers in a manner acceptable to the broadcasters (e.g. subscription free to the end user), there may be no need for a DSO, just an analogue switch-off.

Conclusion

We started this chapter querying the long-term future of DTT because of its spectrum inefficiency and inferior service offering compared to other platforms. But we are ending it noting that any switch-off of DTT before around 2030 is very hard to envisage. Because of DTT's role in PSB, high-level political support would be needed to get rid of the platform or reduce the frequencies it uses. Substantial investments have been made in DTT, and there are currently practical and cost issues associated with replicating DTT's universal coverage and low-cost access proposition.

Attempts to use market mechanisms to determine the future of TV spectrum have so far come to almost nothing. In terms of political and social impact, TV is for many the most important user of the spectrum, making governments very resistant to changing policy. It is also a carefully constructed market with a complex ecosystem where small changes could have profound social effects. But could these problems be overcome? Almost certainly. In all but a handful of countries, there is no political will to even start liberalising this part of the airwaves, and in the one country that tried, the United Kingdom, there was insufficient will to carry it through.

In the long term, there may be better prospects for applying market mechanisms to the broadcast spectrum, and this is discussed in Chapter 23. The US incentive auction is also an important part of this discussion, but as this came later, took a very different approach, and was not complete at the time of writing, we also consider this in Chapter 23.

8
SATELLITE

Of all the discussions in radio spectrum management, few are more vociferous than between the satellite and mobile broadband industries. In one sense, the animosity is puzzling because the mobile and satellite industries are closely linked. Mobile operators use satellite for backhaul*, and the satellite industry is keen to enhance its mobile services.

At the root of the argument, the satellite industry feels threatened by a newer arrival to the wireless world with designs on its spectrum and a supposedly cavalier attitude to the risk of interference. Some within the mobile world see the satellite industry as cumbersome and unresponsive, and believe it puts dogmatic technical concerns above innovation.

One of the many reasons for this division is that the two industries gained their spectrum in different ways. The satellite industry acquired much of its spectrum holdings before serious attempts were made at liberalising spectrum allocation. For them, attempts to liberalise spectrum have been seen as threats, rather than opportunities. In contrast, a substantial proportion of mobile spectrum has been acquired through liberalised means – in the auctions which have become the dominant assignment method since the late 1990s. In short, the satellite industry believes they are the incumbents and have spent decades building up infrastructure while the mobile industry counters that, unlike the satellite industry, they pay for their spectrum. These two sides perhaps point to the world of difference before and after market liberalisation.

Only one country, the United Kingdom, has attempted to apply market mechanisms to the satellite sector. These attracted huge opposition and were greatly scaled back. This chapter considers the lessons

---

* Backhaul is the link between the mobile base station and the mobile operator's computer networks, which process the calls. This is typically via cable, microwave link or satellite.

that can be learnt from that process, as well as liberalisation initiatives driven by individual satellite operators. But we start by examining how satellite spectrum is currently managed.

How the Satellite Industry Uses Spectrum

In 1957, the world's first orbiting satellite, Sputnik, used frequencies of 20.005 and 40.002 MHz to send a series of otherwise meaningless beeps back to earth. Modern satellites handle huge amounts of data and use a large range of frequencies to do so. Allocations for satellite services in the International Telecommunication Union Radiocommunication (ITU-R) sector's Radio Regulations range from frequencies as low as 7 MHz and as high as 275 GHz – far higher than any current terrestrial technology is capable of using.

Satellites can be found in low earth orbit, which is typically between 240 and 800 km above the Earth's surface, and in geostationary orbit (GEO), which can be as high as 35,000 km. At low orbits, satellites criss-cross the Earth at speeds of 27,000 km/h, whereas GEO satellites rest above the same point above the earth, although they are actually travelling around 11,000 km/h through space. Between the two of these is medium earth orbit (MEO), in which satellites typically perform two orbits of the Earth each day.

Table 8.1 provides a summary of some of the most important uses of spectrum by the satellite industry.

Note that Table 8.1 is by no means exhaustive and does not take into account many satellite services, constellations and companies that do not fit into this generalised summary. For example, Thuraya offers mobile communication services but does so through terrestrial Global System for Mobile roaming agreements, and two GEO satellites that cover Europe and the majority of Africa and Asia. Many countries also have a national satellite company that may operate a GEO satellite. Different regions also depend on different bands to varying extents. For example, C-band is heavily used in tropical regions as these signals do not suffer from rain-fade, but other regions are less dependent on this band.

Satellites are also used for earth exploration, and meteorological services, as well as providing maritime and mobile connectivity, imaging and various types of location services.

Table 8.1 Summary of Satellite Applications

ORBIT	FREQUENCY	SERVICE	IMPORTANT CONSTELLATIONS
Low earth orbit	L-band (1.5–1.7 GHz) and S-band (1.9–2.7 GHz)	Mobile Satellite Services	Iridium NEXT (66 satellites) (under construction) Globalstar (32 satellites) (second generation under construction)
Medium earth orbit (MEO)	L-band (1.5–1.7 GHz)	Global Navigation Satellite System (GNSS) A GNSS constellation requires 24 satellites and is typically operated by the public sector	United States: GPS Russia: GLONASS European Union: GALILEO (under construction) China: BeiDou (under construction)
	Ka-band (17.3–30 GHz)	A number of companies are planning broadband satellite services using constellations of MEO satellites	O3B (under construction)
Geostationary orbit	C-band (3.4–7 GHz) Ku-band (10.7–14.5 GHz) Ka-band (17.3–30 GHz)	Fixed Satellite Services (FSS). These are support services such as mobile backhaul and broadcasting and are typically operated by commercial companies	SES (53 satellites) Eutelsat (35 satellites) Intelsat (28 satellites) Inmarsat BGAN (4 satellites) Chinasat (13 satellites) Astra (16 satellites)

National administrations can exempt themselves from these obligations through the use of footnotes which are placed in the ITU's Radio Regulations. Most notably, at WRC-07 around 80 countries used footnotes to identify the 3.4–3.6 GHz band for mobile broadband, which is referred to by the ITU as IMT. Every single EU member state can be found on that list, aside from Luxembourg, home of satellite industry behemoth, SES. In the European Union, the band which is identified for IMT is 3.4–3.8 GHz, which is wider than the 3.4–3.6 GHz band identified via footnote in the Radio Regulations. In countries which seek to use this band, it is shared between mobile broadband and the satellite industry using terrestrial mobile exclusion zones around satellite Earth stations and other interference mitigation techniques.

A single satellite can cover up to a third of the Earth's surface, and the average lifetime of a satellite is 15 years. After a launch, the 'facts in the sky' are impossible to change and affect huge areas that have little to do with national boundaries. Careful international coordination is seen as crucial so that the thousands of satellites that orbit the earth can use the spectrum efficiently and without interference.

The result is that national administrations can only regulate the domestic use of satellite *services*, rather than the frequencies that they rely upon. To acquire the right to use a frequency, operators need to apply to national administrations to make requests for satellite frequencies on their behalf at the ITU-R level. These requests are known as Advance Publication Information, and if its requested slots and frequencies are approved, then the ITU-R will issue a notification to formalise the allocation. Information about frequencies is contained on the Master International Frequency Register.

Spectrum for satellite operators is acquired at the ITU-R level through administrative means. Acquiring spectrum is also a political exercise as satellite operators need to persuade national administrations to sponsor applications for satellite filings, and for the services to be licensed domestically. This tends to see the host country of the operator apply on their behalf. SES applies through Luxembourg, Nigcomsat through Nigeria and Inmarsat through the United Kingdom.

The satellite industry argues that the economic arguments behind spectrum liberalisation are not applicable to the satellite industry because of the way it uses spectrum. Nevertheless, the mobile broadband community insists that these arguments only conceal the satellite industry's inefficient use of spectrum. They often illustrate their point by relating the history of the S-band in Europe.

The S-Band

Satellite systems have long used the S-band. For example, in the 1960s, NASA's Apollo program used the Unified S-band system, which was based on 2025–2110 MHz for uplink and 2200–2290 MHz for downlink.

In February 2007, the European Commission decided to allocate the S-band to mobile satellite services (MSS). Specifically, it authorised

2×30 MHz of spectrum at 1980–2010 and 2170–2200 MHz. An agreement was reached with the European Parliament to assign the licences in a beauty contest, where the winners could have a maximum of 2×15 MHz – half the available spectrum.

Note that in the United States, the Federal Communications Commission (FCC) plans to auction spectrum known as the AWS-3 and AWS-4 bands, which fully impinge on the MSS allocations.

The licences were awarded to Inmarsat and Solaris Mobile. The former is a British company that was born out of the International Maritime Satellite Organization's privatization in 1999 and runs a global network of MSS. Solaris was a joint venture formed by Eutelsat and SES to take up the opportunity of providing MSS using the S-band.

Solaris Mobile's S-band payload was launched a few months behind schedule on 6 April 2009 on Eutelsat's W2A satellite. On 14 May 2009, the same day that the European Commission granted the two S-band licences, Eutelset and SES reported an 'anomaly' in the functioning of the payload. There had been a fault with the payload's coverage and power. Writing at the close of 2014, Solaris Mobile's existing service remains largely uncommercialised, and Inmarsat had yet to launch a payload to use its frequencies, though it has announced plans to do so. The European Commission has been able to do little, except occasionally appeal for action from the relevant parties.

While this 2×30 MHz in the S-band appears neglected by the satellite industry, the 2.1 GHz mobile band (Figure 8.1) which is directly adjacent to the S-band MSS frequencies is well used and was sold for extraordinarily high prices in the infamous 3G auctions in Europe in 2000. In Germany this raised DM 98.8 billion, or €50.5 billion and in the United Kingdom, £18.5 billion was raised.

It is difficult to overestimate the importance of the 2.1 GHz band, known in the standards world as 3GPP band 1, to the worldwide success of 3G. According to the telecoms statistics group, GSA, the band accounts for the majority of the world's 547 commercially launched 3G networks across 205 countries. Inmarsat and Solaris are sitting on 2×30 MHz of adjacent spectrum without paying the regulator – in this case the European Commission – anything, and they are not using it. Given the link between economic growth and broadband growth, there have been critics of such fallow use of spectrum.

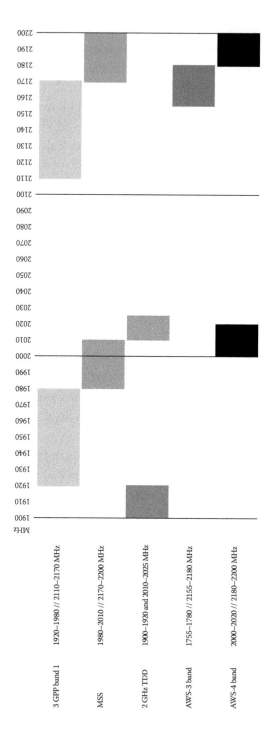

Figure 8.1 S-band spectrum allocations.

Unused Spectrum

While the EC continued its persuasion campaign, both SES's and Eutelsat's 2009 annual reports assured investors that they remain 'confident of Solaris Mobile's ability to meet the commitments made to the European Commission in terms of timing and scope of the mobile services'. A year later, this confidence had changed to commitment to 'establishing an economically viable business through the commercialisation of the awarded S-band frequencies'. This phrase was repeated for the next three annual reports despite the fact that in November 2010, Solaris Mobile's partners decided to cancel €120 million of shares, thereby reducing its share capital to €30 million. Nevertheless, in 2011, Solaris promised the Italian media group Class Editori that it would launch a satellite in 2012, so it could offer a hybrid satellite and terrestrial network.

Solaris Mobile was brought for €24.8 million by the US satellite service provider, EchoStar, in January 2014. According to EchoStar's annual report, Solaris Mobile 'lacked certain inputs and processes that would be necessary to be considered a business'.

EchoStar's acquisition of the company prompted speculation that it was mostly a ruse to acquire the 2 GHz spectrum, and then to launch a terrestrial network on the spectrum. This is legally possible because satellite services in the European Union are permitted to have a terrestrial component. Indeed, Solaris's original business plan included the use of terrestrial communications.

However, EchoStar's own 2014 regulatory filing shows that it seeks to use the spectrum to offer satellite services in Europe.

Although Inmarsat has yet to launch an S-band satellite, it has promised that it will do so since its 2011 annual report[*] and in June 2014 the company released a pamphlet called 'Inmarsat S-band services'. The pamphlet said that the terrestrial network will use the S-band but that the feeder links which connect the satellite access stations to the satellites shall probably use the Ka-band, with the

---

[*] The report reads, 'we are also exploring the development of hybrid opportunities through our S-band satellite programmes in Europe. We believe that over the long term, interference issues can be managed and hybrid systems can be effectively deployed. We also believe hybrid systems could offer attractive new services globally'.

possible use of the L-band. According to sources in the company, Inmarsat plans to primarily use the spectrum to provide in-flight passenger Wi-Fi, in a system known as 'Aero Complementary Ground Component'.

The satellite industry's S-band authorisations in Europe show how different rules have been applied to adjoining spectrum because it is being used by different industries. The policy issue is whether this piece of satellite spectrum has been inefficiently used because, unlike the neighbouring mobile spectrum, it was not assigned using market mechanisms. Writing in 2015, the EU's administrative approach has not achieved its objectives, although bad luck – in the shape of the failure of the Solaris satellite – has certainly played a part. However, the longer the S-band spectrum remains unused, the weaker the underlying policy appears.

The S-band situation also underlies the complexity of merging two industries – satellite and mobile. In the United States, the S-band was essentially given to companies looking to provide MSS with what is known as an Ancillary Terrestrial Component (ATC). This is known in Europe as a Complimentary Ground Component. In short, this is a hybrid satellite and terrestrial network that is designed to use satellite connectivity when outside the range of the terrestrial mobile network. Mobile operators feel that giving such services spectrum for free is unfair. However, the nature of raising the funding and launching a satellite costing millions of dollars leaves the MSS community feeling justified. That said, the work of US companies such as Terrestar and Sky Terra has been limited in terms of what connectivity they are providing to the rural parts of the United States.

Unfilled Satellite Slots

Of course, European MSS services in S-band are just one example of how the satellite industry uses the spectrum. Another aspect of this discussion is the issue of unfilled satellite slots, known as 'paper satellites'. Administrations make applications for satellite slots, and when granted these are reserved for several years. However, if the satellite is never launched, or moves out of its orbit due to a technical problem, the slot then becomes vacant and, therefore, arguably, an inefficient use of spectrum.

Since 1998, the ITU has been trying to deal with this problem, and while it has had a great deal of success, it remains a live issue and will be discussed again at the 2015 World Radio Conference. The ITU has increased fees to try to tackle this issue, but these are nothing like the level paid at auction by terrestrial mobile operators. From the liberalisers' point of view, paying a market rate for satellite spectrum would ensure that these slots are used more efficiently.

The next part of this chapter examines the two attempts to liberalise the sector, one from a regulator and another from the market itself.

The United Kingdom Attempts to Liberalise Satellite Spectrum

The United Kingdom is the only country that has attempted to apply market mechanisms to satellite spectrum. The main thrust of their policy intervention was rejected by the industry and only implemented much later in a highly restricted form. The debate that ensued does much to highlight the specific nature of the satellite industry and the difficulties of taking a liberalised approach.

The liberalisation of the satellite sector was to be based on Recognised Spectrum Access (RSA), a concept introduced into UK law in the Communications Act, 2003. The idea behind RSA was that it could allow regulation of spectrum use, even though conventional licensing is impossible for technical or legal reasons. For example, spectrum use by radio astronomers could not be licensed because the radio transmissions that they seek to detect are from distant galaxies and pay no respect to rules laid down by regulators. Nevertheless, radio astronomers would benefit from legal recognition that the spectrum cannot be subject to harmful interference. The RSA was intended to be a voluntary scheme, and the emphasis was on the application of RSA to spectrum used by the public sector.

Like radio astronomy, Ofcom has no legal or practical powers over satellite transmissions in space (other than for those satellites registered in the United Kingdom), but Ofcom felt that the satellite industry and other spectrum users would benefit if the space-to-earth transmissions were subject to RSA.

In 2002, Ofcom published an independent Review of Spectrum Management, written by the regulatory economist and academic Professor Martin Cave. The report called for all spectrum users to be charged so that they are incentivised to use the spectrum efficiently. Although the document did not mention RSA, it did advocate charging the satellite industry a price that would be based on the opportunity cost of using their spectrum for terrestrial applications.

In April 2005, Ofcom issued a consultation on applying RSA to the satellite industry. Ofcom's satellite consultative committee was vigorously against this, as was the satellite industry.

In off-the-record briefings to the press, representatives of large satellite companies described RSA as a 'protection racket'. They gave two arguments. First, that it could lead to 'unsustainably high fees for satellite operators', particularly if other countries followed Ofcom's lead and applied RSA. Second, they argued that the potential creation of a secondary spectrum market could obliterate the concept of internationally harmonised satellite spectrum. This would make satellite services unviable. Similarly, the European Satellite Operators Association (ESOA), argued that the European single market was undermined by the proposals because it added a regulatory barrier in the country of reception. Ofcom argued that it did not break European law because RSA was voluntary. The satellite community argued that in order to receive protection against interference, the industry did not face a truly free choice.

More fundamentally, it was argued that RSA did not inherently offer protection from interference and that other spectrum management tools would be more appropriate.

The ESOA argued that there was no real demand for this spectrum from any viable other service, and said that they believed RSA, opportunity cost, and the desire to create economic value was a dogma for Ofcom.

In December 2005, Ofcom published another independent review by Martin Cave, the highly influential 'Spectrum Audit' discussed in Chapter 6. This focussed on public sector spectrum, but also recommended that Ofcom should consider the use of RSA for receive-only satellite earth stations in the C-band (3.6–4.2 GHz) and in the Ku-band. Cave wrote that not only could Ofcom charge satellite operators for using this spectrum but it would also 'make satellite use more transparent, give satellite operators a degree of statutory recognition, and assist planning'.

Although Ofcom pressed ahead with its plan to apply RSA to radio astronomy in 2006, the subject of applying RSA to commercial satellite users of spectrum was left dormant for half a decade. But in June 2010, Ofcom issued a consultation on introducing RSA to the C-band, as well as other spectrum used for the meteorological satellite service (Metsat) in the 1690–1710 MHz and 7750–7850 MHz bands. In the associated information, Ofcom wrote that the proposal followed a request for formal recognition in Ofcom's spectrum management and planning process to limit the possibility of interference from terrestrial services in these bands. The majority of stakeholders were critical about the proposal in their consultation inputs, particularly on Ofcom's proposal to base the fees on administered incentive pricing (AIP), rather than fees based on recovering Ofcom's costs in protecting the spectrum from interference.

Nevertheless, in May 2011, Ofcom announced that it was going to establish RSA for Receive-Only Earth Stations (ROESs), and was going to apply charges on users for the use of this spectrum that would be based on the level of protection a user could expect to receive from Ofcom for the band, and AIP. The decision says that companies who wished to avoid paying the fees were not obliged to be subject to RSA. Since then, RSA has been applied to ROESs that use the 1690–1710 MHz, 3600–4200 MHz and 7750–7850 MHz bands.

In September 2014, Ofcom proposed to extend the scheme to ROESs that use the 7850–7900 MHz and 25.5–26.5 GHz bands. It said that the new proposals follow at least two new ROESs that are planned and could suffer interference from fixed links in the same band.

Over 10 years after RSA was first proposed, it is now applied to five ROESs and six radio astronomy sites. No spectrum trades have been enabled due to RSA. The fundamental problem is that satellite services cover regions, not countries. If satellite operators were to face spectrum charges in the United Kingdom but not in other countries, they would move elsewhere. That problem could be avoided if spectrum charges were applied internationally, but there is little appetite for this and the satellite operator community would be likely to oppose such a move very strongly, limiting its chance of success. One argument used by the satellite community is that full liberalisation – if applied internationally – could see the harmonisation of satellite bands eroded by trading.

The United Kingdom has probably gone as far as possible by applying market mechanisms to the receive-only aspect of satellite transmissions, concentrating on the C-band where there is substantial interest from the mobile community and therefore a greater justification for a market approach to regulate supply and demand.

Operator-Driven Proposals

Satellite operators' opposition to liberalisation is based partly on their concerns about interference. They argue that not having exclusive, harmonised bands can lead to problems, and there is some support for this argument in the story of LightSquared in the United States. This company tried to use a satellite band for mobile services but had to abandon the plan because of the interference it would cause to neighbouring satellite Global Positioning System (GPS) services.

In March 2010, Harbinger Capital, a hedge fund, acquired 2 × 35 MHz of spectrum in the L-band when it brought the satellite operator, SkyTerra. Under its enigmatic and litigious owner, Philip Farcone, the company signed an 8-year, $7.2 billion contract with NSN (then Nokia Siemens Networks) to build a Long Term Evolution (LTE) network that would use the 1.5 GHz band spectrum. This network was planned to sell wholesale capacity for mobile networks and was known as LightSquared.

He was able to do this because the FCC had introduced an aspect to the MSS licences that allows companies to build an ATC. This meant that an operator could run a terrestrial-based standard LTE network using an MSS licence. For LightSquared, this meant a network of 40,000 terrestrial base stations and two satellites that could communicate directly to a customer's handset. The company's business model focussed on utilising this hybrid network model of urban terrestrial connectivity combined with rural satellite connectivity. Given the focus that all governments have on ensuring that rural areas have equal access to broadband under their Universal Access schemes, the proposals of companies such as LightSquared have generally been deemed attractive.

In January 2011, the FCC went further and gave the company a waiver from a licensing condition that stipulated that all devices using this band would have to be able to communicate with satellites. This meant that the company could use standard LTE devices, which would be much cheaper than developing a hybrid satellite/LTE device. Without the waiver, the company argued that its network would have been uneconomical. After having jumped through the relevant regulatory hoops, LightSquared predicted it could offer mobile services to 260 million subscribers by 2015. It initially appeared that a liberalised spectrum policy framework had succeeded in re-farming spectrum from a failed business model to a more productive use.

But by that summer, LightSquared had come up against serious problems. It had emerged that the original plan would cause substantial interference to adjacent GPS users such as the Department of Defence (DoD) and the Department of Transportation (DoT). The GPS industry and LightSquared had a public row in which LightSquared argued that the GPS equipment was oversensitive to interference.

In an attempt to keep the plan on track, LightSquared submitted an alternative proposal, which would have involved a spectrum swap (1546–1556 MHz with 1670–1680 MHz), a drastic reduction of power levels and a 23 MHz guard band. The GPS industry rejected this proposal, demanding a 34 MHz guard band instead (Figure 8.2).

The following February, the US public sector spectrum regulator, the National Telecommunications and Information Administration (NTIA), wrote a letter to the FCC in which it concluded that interference problems from LightSquared's proposed network were unresolvable. These findings were disputed by LightSquared, but there was nothing it could do, its business model was broken.

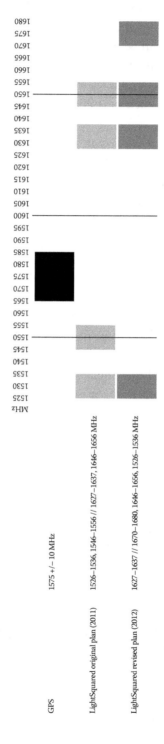

Figure 8.2 LightSquared frequency plans, showing the move away from GPS spectrum.

Conclusion

Spectrum policy for satellites is still largely based on a command-and-control method. This is because satellite deployments are expensive, have a large up-front investment, take a long time to plan and are international services operated by global companies. However, some argue that this has led to inefficient use of spectrum, and possibly some unjust enrichment. Attempts have been made to liberalise the spectrum by both regulators, and some entrepreneurial holders of satellite spectrum. So far, success has been limited.* The reliance of the mobile broadband industry on satellite to provide backhaul may be waning with the growth of fibre networks, but the creative use of MSS will influence global connectivity for many years to come.

---

* At the time of writing, a US company, Globalstar, was trying to convert its 2.4 GHz MSS spectrum to be used as an extension of the unlicensed 2.4 GHz Wi-Fi band. This had yet to receive FCC approval.

9

ULTRA WIDE BAND

Ultra wide band (UWB) is an example of an underlay network (i.e. it operates at a very low power across a wide frequency range, 'underneath' other radio technologies). The theory is that because its signals are transient and have little energy, they do not cause interference to the existing users of the spectrum, and therefore should allow spectrum to be efficiently reused.

All radio signals are subject to a small level of interference. This comes from a wide range of sources, including other electronic devices, thunderstorms and even from the Big Bang at the start of the universe. This background level of interference is known as the noise floor, and in general terms, UWB networks are meant to operate beneath this noise floor, while having special characteristics which enable suitably designed receivers to pick up their transmissions.

From 2000 onwards, UWB was promoted as an upcoming new technology which could produce high data transfer rates and had a variety of applications, including replacing Ethernet cables in office networks and superseding USB cables on laptops and PCs. Liberalisers were keen to accommodate UWB as it offered the chance to wring more value from the spectrum by sharing with existing users. Administrations spent a great deal of time drawing up a regulatory framework for UWB but by 2008 as a mass market product it was doomed: WiQuest, the biggest manufacturer of USB products, closed down. The technology had failed to deliver on its initial promise and only continued to be used for a few niche applications.

Principles of an Underlay Network

Every radio system operates in an environment where its performance is limited by the noise and interference it experiences on the frequency on which it is operating. It is also possible to spread out information over a much wider amount of spectrum than is strictly necessary to convey it, thus making it appear as noise across a range of frequencies rather than as interference on a specific frequency. When information is spread out in this fashion, the amount of information on any given frequency is reduced. If the information is spread out over a wide enough range of frequencies, the amount of signal that is present on any given frequency can be reduced to such a level that it is lower than the 'normal' amount of noise or interference, effectively making its impact to any other system using that frequency virtually imperceptible. This is the principle of an underlay network: information is spread out over such a wide range of frequencies that its impact to other radio systems is imperceptible. It sits 'underneath' other radio services.

The idea of such a system dates back to the 1960s and 1970s through the work of pioneers such as Dr. Henning Harmuth and Dr. Gerald Ross. It was realised that generating a tiny impulse of electricity, with a duration of say, 1 ns, would produce a radio signal that would occupy 1 GHz of spectrum. Shorter impulses occupy even more spectrum. These impulses can contain useful amounts of energy yet spread out across so much of the radio spectrum that the amount of interference they cause to other radio signals is negligible.

To use an example, a walkie-talkie radio might transmit with a power of 5 W and use a 12.5 kHz radio channel (i.e. the information is spread out over 12.5 kHz). We will compare this to a 5 mW impulse transmitter (designed to offer only short-range communications). We will assume that the walkie-talkie we are trying to receive is 1 km away and that the impulse transmitter is 10 m away and that the frequency used by the walkie-talkie is 450 MHz (Table 9.1).

In this example, if the impulse signal occupies 12.5 MHz of spectrum, the interfering signal it would cause at the walkie-talkie receiver will be 100 times smaller than the wanted signal from the walkie-talkie transmitter. This is insufficient to cause a problem. If the impulse signal is spread out a further 100 times such that it occupies 1.25 GHz of spectrum (a value more representative of proposed systems), the interference caused to the walkie-talkie is now 10,000 times smaller than the wanted signal, far below the point where it would cause any problem at all. This is the case even

Table 9.1   Relative Power Levels of a UWB Transmitter to a Walkie-Talkie

TRANSMITTER POWER	SPECTRUM OCCUPIED	PATH LOSS (FREE SPACE)	POWER RECEIVED (IN 12.5 KHZ BANDWIDTH)	RELATIVE LEVEL
5 W (walkie-talkie)	12.5 kHz	1-km path gives 85-dB loss	−48 dBm (16 nW)	
5 mW (impulse device)	12.5 MHz	10-m path gives 45-dB loss	−68 dBm (0.16 nW)	100 times smaller
5 mW	1.25 GHz	10-m path gives 45-dB loss	−88 dBm (0.0016 nW)	10,000 times smaller

though the walkie-talkie is 1 km away, but the impulse device is only 10 m away. Combining this tiny interfering signal with the very short duration of the impulse, it can be seen that this technique could share spectrum with traditional radio systems without causing them harmful interference.

Unlike traditional radio receivers which listen for a signal on a specific frequency, receivers for this impulse system would instead listen across the whole of the spectrum. Traditional radio transmissions are of such a relatively long duration compared to the impulses that they are invisible to the impulse receiver, and thus it is possible to distinguish the impulses even in the presence of traditional radio signals.

Using impulses is only one way of generating a signal that is spread out over a wide range of frequencies. The general name for these technologies is ultra wide band reflecting the large amounts of spectrum they occupy.

Technological Hurdles

Such a technology presents a number of engineering problems. First, the frequency range over which a normal radio antenna functions is typically restricted by the physics of the antenna. Even the most 'broadband' antennas typically operate over a frequency range that spans no more than a factor of 10 (known as a decade, e.g. 30–300 MHz), and most have much smaller bandwidths than this. Second, generating such high-speed impulses requires very-high-speed electronics (i.e. devices capable of switching on and off within a nanosecond or less), which were not easily manufactured when the concept was first developed. Receiving the impulses requires equally high-speed electronics. As such, the idea remained a largely theoretical concept, or one confined largely to research laboratories, until the late 1980s at which point many of the engineering problems were able to be overcome through developments in semiconductor electronics.

Uses of the Technology

Once it became technically feasible to generate such signals, the next step was to agree to a regulatory framework in which they could operate. In the United States, the Part 15 regulations provide a threshold for permitted emissions from an electronic device without the need for a specific licence. These regulations form the genesis of contemporary discussions of spectrum sharing policy.

The UWB proponents argued that as long as their technology produced emissions below the thresholds set in the Part 15 regulations, they would require no individual licence and thus that no specific regulation was necessary in order for them to operate. They argued that underlay networks provided a means to share spectrum between different technologies and that this was a highly efficient use of the spectrum.

While it is clear that the UWB technique should not cause harmful interference to other radio systems, regulators and other radio users remained concerned. In particular, they were worried about the combined effect of multiple devices, the effect they may have on signals that are already very weak and the similarity of UWB impulses to radar pulses.

A single UWB transmitter should produce a series of pulses spread out across a wide range of frequencies such that it causes little or no interference to traditional, narrowband radio spectrum users. But regulators were concerned about the cumulative effect of multiple devices. If two impulses coincide (at a receiver), then the amount of interference received could double. At what point would the impact of multiple devices combine to begin to cause interference? A lot depends on the density of devices and their proximity to the system to which they could cause interference. In order to arrive at a result, it is necessary to make assumptions and hypothesise use cases. Given the inherently statistical nature of the results of any such calculations, they do not provide the kind of regulatory certainty that spectrum managers like.

Another concern was the impact of UWB transmissions on the reception of signals that are already very weak. Of particular concern were the signals from GPS satellites, which on the surface of the Earth are barely above the noise floor (technically speaking they are at a level of -155 dBW and the equivalent noise level at room temperature is -141 dBW, meaning that a GPS signal is already typically 14 dB or 30 times weaker than noise). Any increase in the level of interference could therefore have a drastic impact, but much depends on the density of UWB devices and their proximity to GPS receivers.

A final concern was that the impulse-like transmissions from UWB devices were similar in nature to the transmissions from radars, and that a radar might interpret the reception of UWB signals as reflections from a target, producing false responses on the radar display.

Regulating UWB

In order to try and overcome these concerns, it was decided in 2002 by the FCC that UWB equipment could not operate under the Part 15 regulations and that specific restrictions on emissions would be needed to ensure protection of the services concerned. Noting the concerns over the potential interference from UWB devices, the FCC order[*] states:

> This has been an unusually controversial proceeding involving a variety of UWB advocates and opponents. These parties have been unable to agree on the emission levels necessary to protect Government-operated, safety-of-life and commercial radio systems from harmful interference. It is our belief that the standards […] are extremely conservative.

These FCC orders permit the operation of UWB devices in the frequency range of 3.1–10.6 GHz at the standard Part 15 emission

---

[*] Federal Communications Commission. (2002). Revision of Part 15 of the Commission's Rules Regarding Ultra-Wideband Transmission Systems (https://apps.fcc.gov/edocs_public/attachmatch/FCC-02-48A1.pdf).

levels but add additional restrictions on other frequencies. The FCC mask for UWB emissions is shown in Figure 9.1.

The following are notable:

- The permitted level of emissions for UWB devices is below the Part 15 levels (i.e. below 3.1 GHz).
- A reduction in emissions is required across a wide range of frequencies, in particular between 960 and 1610 MHz, which includes many aeronautical systems (e.g. radars) and GPS frequencies.
- The resulting emission mask for UWB devices is not flat (i.e. emissions cannot be equal at all frequencies across the range in which devices are permitted to operate).

This latter point may not initially appear on the surface to be overly restrictive; however, it punches a hole in one of the fundamental operating principles of UWB devices. An impulse spreads power *evenly* across a range of frequencies – it cannot easily be adapted to produce different levels of signal at different frequencies. The only way in which such a frequency response can be produced is through the generation of complex waveforms or the use of clever filters, and in doing so, one of the key premises of UWB is lost – the simplicity of the pulsed transmission and reception.

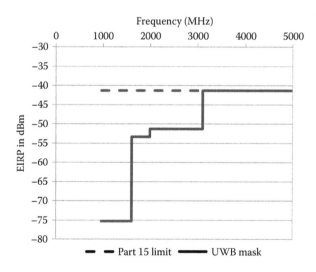

Figure 9.1 FCC UWB emission mask.

The FCC is not the only regulator to produce an emission mask for UWB devices, and permitted powers and frequencies vary between countries and regions. Most UWB devices therefore use orthogonal frequency division multiplexing to produce a number of carriers, spread across the permitted frequency operating range with the power levels of each carrier tailored to the emission mask that they have to meet in each of the different regulatory regimes. Thus the unique advantages (in particular the one of simplicity) of UWB devices are lost to the need to support differing regulatory approaches.

What Have We Learnt?

The tight regulation of UWB, whether justified or not, played a part in its demise. The restrictions made it difficult to deliver the promised data rates of 480 Mbps, and the need to establish international standards and internationally consistent regulatory approaches slowed down the technology's development. In the meantime standardisation efforts to increase the speed of Wi-Fi were proceeding apace, and these rapidly achieved critical mass in an already crowded market, inflicting a killer blow on the staggering UWB manufacturers.[*] The manufacturers were also criticised for offering products where there was no pressing market demand. For example, is there a compelling need for wireless USB? The cables are extremely cheap and having one connected to your computer is hardly a huge inconvenience. Many peripherals such as mice, keyboards and printers successfully use other wireless technologies to replace USB leads without the need for UWB.

What do UWB's difficulties tell us about spectrum liberalisation? First, existing users are reluctant to share their spectrum. Most spectrum users believe they have exclusive access, even though enlightened regulators devise licensing agreements, which make it clear this is not the case. There is no incentive for existing users to grant access, so it can take many years for regulators to talk them round. UWB's business model depended on free spectrum: existing users had nothing to gain by granting access so the permitted 'interference levels'

---

[*] The following tells the story of the last years of UWB: Fleishman, G. (2008). Ultrawideband: Another One Bites the Dust. *Artstechnica* (http://arstechnica.com/uncategorized/2008/11/ultrawideband-another-one-bites-the-dust/).

were so low that the technology was compromised from the start. If existing users had an incentive to share perhaps the story would have been different.

Second, the need for international regulatory agreements to ensure a big enough market for the technology delayed the process even further. This need for international harmonisation is a common problem for spectrum liberalisation. It means there is very little liberalisation in satellite and the public sector and we will see it cropping up again later in the book when we discuss in Chapters 13 and 14 the need for more mobile spectrum.

Underlay networks have the potential to open up the airwaves to more users and more uses, a key goal of spectrum liberalisation. However, as we have seen with UWB, they are difficult to implement in practice. Existing users' reluctance to share spectrum is one reason for this, but there are other reasons including the failure to offer 'must-have' products, which are clearly better or cheaper than other technologies.

At some stage in the future, market conditions may change, regulatory conditions may become more favourable and underlay technologies may have their day. Writing in 2015, UWB was successful in only a few niche areas which take advantage of its radar-like characteristics. The resolution (accuracy) of radar is largely driven by the bandwidth it occupies. Radar occupying a very wide range of frequencies therefore has the potential to be very accurate. Using UWB transmitters as radars therefore permits accurate measurement of distance and it has been found that high-power UWB transmitters can be used as accurate building penetration radars and are often used by the fire service to develop a picture of the insides of a burning building. In addition, radio frequency identification tags attached to stock in a warehouse can be tracked accurately with UWB radars, finding the items on particular shelves or in particular boxes.

10
WI-FI AND THE
SPECTRUM COMMONS

It is now becoming almost more important for cafes to offer free Wi-Fi than coffee. This is because for many end-users, Wi-Fi *is* the Internet. But Wi-Fi really is only a family of technologies that uses a small part of the radio spectrum, typically just 83 MHz in the 2.4 GHz band and around 400 MHz in the 5 GHz band.

These bands, and in future the 60 GHz band, stand in stark contrast to other heavily used bands – they are a commons. There are no individual property rights in these bands, and they offer users no protection against interference. Governments set power limits to curb the extent of any interference but no single company has a licence for this spectrum. These are licence-exempt (also known as unlicensed) bands,* and these relatively small bands contain all the traffic carried by Wi-Fi, Bluetooth and a host of emerging technologies.

Where does unlicensed spectrum stand in our discussion of spectrum liberalisation? There is no single answer: liberalisers vary significantly in their enthusiasm for the unlicensed approach. At the negative end of the scale we have economists such as Thomas Hazlett, who think it is most efficient to sell spectrum via market mechanisms such as auctions. At the opposite extreme, following the take-off of Wi-Fi in the United States around 2000, a common call was to treat all spectrum as a commons. This view has been labelled 'California dreaming' by economist Gerard Pogorel.†

---

* Note that under the International Telecommunication Union (ITU) Radio Regulations, no spectrum can be used without a licence. Most administrations issue a general authorisation that permits devices such as Wi-Fi to occupy specific bands with specific technical characteristics, effectively a single licence or authorisation for all compliant devices.

† See Pogorel, G. (2007). Opinion: The Nine Regimes of Spectrum Management. *PolicyTracker* (https://www.policytracker.com/headlines/opinion-the-nine-regimes-of-spectrum-management).

Most liberalisers sit somewhere in the middle. Cave, Doyle and Webb (2007) argue that spectrum should be unlicensed 'if it is unlikely to be congested'.* Wi-Fi is very low power, can detect nearby networks and share the resource, and has the ability to change channels to avoid interference, so minimising congestion. This means that Wi-Fi is well suited to an unlicensed approach. The authors point out that whether this approach should be extended into other bands is a key area of debate, a point echoed by another key thinker, Lawrence Lessig.

'There is no reason to embrace *either* the market *or* the commons completely', says Lessig, arguing that the best strategy is to embrace both, marking off 'significant chunks of spectrum for sale, while leaving significant chunks of spectrum open in a commons'.† Lessig's liberalising instincts are particularly attracted by the lack of bureaucratic or governmental control over unlicensed spectrum – allowing innovators quick access to spectrum without any political interference.

Unlicensed spectrum also lowers the barriers to entry for new technologies. There is no requirement to buy a spectrum licence, and these can be prohibitively expensive. So a mixture of market approaches, including both licensed and unlicensed spectrum, should stimulate innovation. We regard an element of unlicensed spectrum as being an inherent part of the liberalisation project, while accepting that there is an ongoing debate about how large this element should be.

In this chapter, we look at how Wi-Fi has developed and how its proponents have addressed the arguments mentioned above.

An Incredible Success Story

'Wi-Fi might be the most exciting and important use of unlicensed spectrum in, say, 2000 years', said Federal Communications Commission (FCC) Commissioner Jessica Rosenworcel at a Silicon Valley event in September 2014.

It is easy to see why she said that. According to the Wi-Fi Alliance, in 2013 there were around 2 billion Wi-Fi device shipments

---

* Cave, M., Doyle, C., and Webb, W. (2007). *Essentials of Modern Spectrum Management*. New York, NY: Cambridge University Press, p. 218.
† Lessig, L. (2001). *The Future of Ideas*. New York, NY: Vintage Books, p. 222.

and over 5 million public hotspots worldwide. The annual global economic benefit of licensed-exempt spectrum has been estimated at $270 billion.*

However, none of this would have been possible without a controversial decision by the FCC on 9 May 1985, which allowed unlicensed access to lightly used parts of the radio spectrum for communications, provided the devices use a little-known technology called 'spread spectrum'.

Spread spectrum was developed by the US military in the 1970s as a way to counter potential radio jamming from foreign agencies by spreading the radio waves over more spectrum than actually needed to carry the requisite information. The technique is also handy for preventing interference from more innocuous, but noisy, civil uses of spectrum.

In the 1980s, Michael Marcus, who worked in the FCC, realised that technology could have important applications beyond the military. He managed to persuade his employer to fund research into the emerging technology. By 1984, his team had narrowed the FCC proposals to authorise spread spectrum in three ISM (industrial, scientific and medical) bands in the United States (902–928, 2400–2483 and 5725–5875 MHz). These bands were lightly used, apart from by domestic and commercial microwave ovens in the 2.4 GHz band. The use of the band by microwave ovens had fortuitously discouraged other radio spectrum users from trying to acquire rights for these 'garbage bands'.

Although the 1984 proposal was passed in May 1985, there was a great deal of internal backlash against the proposal and Marcus was moved sideways for 7 years.

In 1988, the Institute for Electrical and Electronics Engineers (IEEE) set up the 802.11 committee to come up with interoperable standards for local area networks (LANs) on this spectrum. The committee published its first interoperable standards 9 years later. These standards were designed to work in the 2.4 GHz (802.11b) and 5.8 GHz bands (802.11a). In 2000, the standard was branded 'Wi-Fi', and is now globally promoted and certified by the 'Wi-Fi Alliance'.

---

* Thanki, R. (2013). *The Case for Permissive Rule-Based Dynamic Spectrum Access*, p. 16 (https://www.policytracker.com/documents/case-for-permissive-rule-based-dynamic-spectrum-access_thanki.pdf).

At the time, it was far from obvious that short-range wireless connectivity needs would be met by technology using an interoperable standard such as Wi-Fi.

The tipping point came when Apple introduced Wi-Fi chips on to their iBook series of computers. Wi-Fi's business success quickly became all but assured, and now it has become difficult to find a networked device that does not have Wi-Fi capability.

At the time of writing, there are 39 variants of the 802.11 standard stretching up and down both the alphabet and radio spectrum. These standards encompass frequencies as high as 60 GHz (802.11ad, also known as Wi-Gig) and as low as 54 MHz (802.11af, also known as White-Fi). Your smartphone or laptop is most likely installed with at least an 802.11n and an 802.11ac capability. The newer standards have much faster download speeds: the 'ac' standard delivers a throughput of 1.3 gigabits per second, compared with just 11 megabits per second for the older 802.11b version.

The Triumph of the Commons

Despite the success of Wi-Fi, by 2005, the FCC's former chief economist, Thomas Hazlett, maintained that the merit of licence-exempt spectrum was unproven. Hazlett told the press that the 'hype' for the commons approach to radio spectrum management was 'peaking'. He insisted that the unlicensed bands constituted 'state property' and that spectrum can in general be more efficiently assigned when it is licensed to private companies.

Critics of unlicensed spectrum often cite the Tragedy of the Commons, a concept that was popularised in a 1968 article in *Science* by Garrett Hardin. The article says that 'freedom in a commons brings ruin to all'.

To make this argument, he cited an 1832 pamphlet that was written in the United Kingdom to promote enclosure of (i.e. taking) common land into private ownership.* This sought to explain why 'the cattle on

---

* Commons were open to anyone to use for purposes such as grazing animals or collecting firewood and were a particularly important resource for the poorest members of agricultural communities. In the United Kingdom, enclosure started in earnest in the sixteenth century and was almost over by 1832. The pamphlet referred to was written by an amateur mathematician, William Forster Lloyd.

a common [are] so puny and stunted', and why 'the common itself [is] so bare-worn'.

Hardin argues that when agricultural land was freely available to all, the inevitable – or according to classicist terminology, tragic – outcome is short-term exploitation that renders the land unusable in the long term.

This is because individuals are incentivised to use the common land to maximise the private value one can accumulate. The costs of doing so, such as decreasing quality of soil, are shared by the whole community, and are in the short term negligible from an individual point of view. Seeing as the community at large has no right to coerce individuals to use the land sustainably, nothing can be done to protect the land from degradation. Further, he argues, as the quality of the soil deteriorates, individuals are incentivised to intensify the use of the land to make up for its declining returns. When all individuals take the same action, the process accelerates exponentially.

A part of Hardin's article dwells on how to deal with scarce areas of natural beauty such as Yosemite National Park, which is to Hardin what beachfront spectrum is to spectrum managers. He argues that the park would be ruined if a Commons access principle was maintained, and so proposed several options, all of which have been tried by spectrum managers:

> We might sell them off as private property. We might keep them as public property, but allocate the right to enter them. The allocation might be on the basis of wealth, by the use of an auction system [spectrum auctions]. It might be on the basis of merit, as defined by some agreed-upon standards [beauty contests]. It might be by lottery [lotteries]. Or it might be on a first-come, first-served basis, administered to long queues [satellite frequency filings]. These, I think, are all the reasonable possibilities. They are all objectionable. But we must choose – or acquiesce in the destruction of the commons that we call our National Parks.

For the record, Yosemite National Park currently charges an entrance fee of $20 per vehicle. The relatively low charge suggests that the infinite demand for use of the park may not have been so infinite as Hardin anticipated.

The Economics of Unlicensed

Far from a tragedy, many argue that Wi-Fi's exploitation of the unlicensed bands has been a triumph. Perhaps, the low barriers to entry associated with a commons approach are a blessing rather than a curse.

This argument was made by the lawyer and activist Lawrence Lessig in his book, *The Future of Ideas*. In it, he argues that complete control over a scarce resource, be it through government regulation or property rights, prevents innovation. Writing at the turn of the millennium, he pointed out that wireless innovators were having to travel to Native American reservations, or even to Togo to be able to innovate with wireless technologies without worrying about spectrum regulations. For him, the more spectrum that can be made available with as few restrictions as possible, the better.

He writes: 'No investor or corporation built the radio spectrum. This resource was given to us pre-built by Mother Nature. Thus, the claim to free access is simply a claim that the government not get in the way of experimentation by innovators.' For him, a commons approach to spectrum access is proven by the success of the Internet, itself a commons because no actor can claim ownership of it.

Lessig also argues that the unlicensed approach allows spectrum sharing, and there is huge potential here because spectrum can be endlessly reused. Concurrent services may interfere with each other, but unlike physical goods, spectrum as a resource does not deteriorate and can be used again and again.

Nevertheless, an unlicensed approach to spectrum management has been described by Hazlett in a 2010 academic paper as the 'Humpty Dumpty approach'.* According to him, this distributes 'a massive number of tiny, overlapping rights that cannot be usefully re-aggregated'.

According to him, the 'rationality supplied by liberal spectrum licences' was proven by the enormous revenues raised by governments in spectrum auctions since the advent of Wi-Fi. In particular,

---

* Hazlett, T. W., and Leo, E. T. (2010). *The Case for Liberal Spectrum Licences: A Technical and Economic Perspective.* George Mason University Law and Economics Research Paper Series 10-19, p. 48. Arlington, VA: George Mason University School of Law.

he pointed out the 700 MHz licences in the United States were worth, according to the market at the 2008 auction, $19 billion.

He points out that mobile operators could have chosen to ignore the auction and have deployed networks using unlicensed spectrum, but they believed that their customers would pay only for networks deployed on expensive licensed spectrum. For him, the cost proves the value.

But for advocates of unlicensed spectrum, value is more complicated than price alone. For example, which is more essential to everyday life – your plasma TV or your Wi-Fi router? The latter, people often reply, even though the TV costs much more. For commons enthusiasts, this proves that the use of unlicensed spectrum has positive externalities. That is to say, the value to society that unlicensed spectrum generates is greater than the price that people pay for it.

In fact, a 'mixed economy' approach to spectrum licensing is being embraced by industry. The mobile sector – which relies on exclusive licensed spectrum – encourages customers to use Wi-Fi for offloading their data consumption from cellular networks. According to the latest edition of Cisco's widely cited Visual Networking Index, 45 per cent of total mobile data traffic in 2013 was offloaded onto the fixed network through Wi-Fi or femtocells (home mobile signal boosters), and that by 2018 more than half of all data downloaded on mobile devices will be offloaded, rather than carried through cellular networks.

If one measures success as the number of shipments and the quantity of data that it has carried, then Wi-Fi has demonstrated a triumph of the Commons.

A Danger of Congestion?

The low power of Wi-Fi and the ability to change channels to avoid interference greatly restricts the amount of interference we might expect from the use of unlicensed spectrum. Most users never experience a problem with their Wi-Fi, but if they do, this is often solved by moving to a channel which is not being used by neighbouring Wi-Fi devices. The use of these channels in 2.4 GHz is shown in Figure 10.1. On some devices, there is also the possibility of moving to the less-used 5 GHz band.

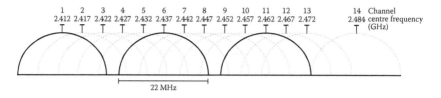

Figure 10.1 Wi-Fi channels used in 2.4 GHz (802.11b/g/n). Note that channel 14 is available only in Japan. (From Michael Gauthier, *Wireless Networking in the Developing World.* Usage licensed under Creative Commons Attribution-ShareAlike 3.0. With permission.)

However, there is still a possibility of congestion in the Wi-Fi bands, and whether this is happening, or will happen, is the subject of considerable debate. Currently, much of this debate focusses on the Internet of Things (IoT), where everyday devices from refrigerators to heart monitors are connected to the Internet and need to send and receive data.

Current experimentation in IoT is based on the same bands as Wi-Fi. And because these bands are unlicensed, there is nothing the Wi-Fi community can do to stop it. Indeed, unlicensed spectrum provides the ideal conditions for innovative and experimental technology such as the IoT.

In the Home Area Network market, Zigbee, HART and Miwi and a host of other standards all use the same 2.4 GHz band as Wi-Fi to send small packets of data such as lighting controls or automatic meter readings. This uncontrolled use of the spectrum could theoretically cause congestion on Wi-Fi networks, although there is little evidence that this has yet been the case.

Whether or not these new technologies will undermine the future success of Wi-Fi is an open question. In 2009, following complaints of Wi-Fi congestion, UK regulator Ofcom commissioned a report by Mass Consultants Ltd. called 'Estimating the Utilisation of Key License-Exempt Spectrum Bands'. It found that congestion in the 2.4 GHz band was not due to a proliferation of Wi-Fi devices, but was instead due to other devices using the same spectrum, particularly analogue video senders. One of the report's authors, Adrian Wagstaff, said at the time that 'No matter how hard I tried to create those city centre conditions, and no matter how hard I pushed it, things just kept on working. There is a very strong set of protocols for Wi-Fi and left to their own devices the networks just keep on working'.

As futurists predict massive deployment of the oT, or even the Internet of Everything, it is certainly possible that legacy Wi-Fi devices that rely on the 2.4 GHz band will become unworkable. On the other hand, not all IoT devices use 2.4 GHz. For example, the Weightless IoT standard was designed to be used in lower frequencies because their superior propagation characteristics mean they could provide the regional coverage needed for applications such as remote reading of electricity meters. Weightless has focussed on TV whitespaces in the 600–800 MHz range and unlicensed portions of the 800 and 900 MHz bands.

However, Weightless, and its competitors will also have to contend with the new Wi-Fi standard, 802.11ah. This new standard is being led by Qualcomm and is specifically designed to support the Internet of Everything's needs: long battery life and 'short bursty data packets'. It is expected to become commercially available in 2015.

Could a Mobile Carrier Rely on Wi-Fi?

Assuming there is clearly nothing inherently 'tragic' about the Commons, why do mobile operators focus on using licensed spectrum?

Some in the mobile industry argue that this is because one needs licensed spectrum to build a cellular network that is private and managed. They argue that Wi-Fi really owes its success to the common practice among fixed-line operators of giving home customers free routers that are equipped with Wi-Fi. So while Wi-Fi may be extraordinarily successful in private places, public Wi-Fi is an unproven concept. By extension, these actors also dismiss claims that more spectrum should be allocated on an unlicensed basis.

However, developments known as Hotspot 2.0 could change that. This technology is based on the passpoint standard. This is a new standard, also certified by the Wi-Fi Alliance, which automates the log-in and authentication process that a user usually needs to go through each time the user accesses a public Wi-Fi network. This technology is allowing firms like Boingo, The Cloud and BT Openzone to provide theoretically seamless but still nomadic Internet access.

Following its acquisition of 2.6 GHz band spectrum in the UK's 4G auction in 2013, BT promised to use these latest Wi-Fi technologies to create what it called an 'inside out network' in the United Kingdom.

This is a mobile network using a combination of BT's newly acquired spectrum and Wi-Fi to turn each of its large networks of home broadband DSL connections into a mini-base station. Areas out of range of an existing hotspot would be covered by BT's existing Mobile Virtual Network Operator or MVNO arrangement. This plan indicates that some sections of the industry think the next stage of Wi-Fi development could have a serious impact on mobile business models.

Or perhaps the real lesson for mobile networks is that their customers are often nomadic rather than mobile. Most people download or upload data while at home or at a place of work. Notwithstanding some Voice over IP technologies such as Skype, nomadic connections do not need the assured, interference-free characteristics of a cellular phone call on a licensed network. Most people's needs, most of the time, are more than adequately met by the fluctuating download speeds present in Wi-Fi networks. In these situations, it makes little sense to send packages a couple of kilometres through flora, fauna, and walls to an expensive base station when a DSL connection that has already been paid for is sitting within walls a couple of metres away.

The increasing reliance of the mobile industry on Wi-Fi was shown in September 2014 with the launch of Apple's iPhone 6, which supports 'Wi-Fi calling'. In the United States, the only company that supports the technology is T-Mobile, which has much less licensed spectrum than its competitors and less advanced plans for Voice over LTE.

Conclusions

Wi-Fi's enormous successes have proven that the commons model of spectrum management can produce impressive results, at least for certain kinds of applications. What were once considered as Garbage Bands now host the majority of wireless communication, if measured by the quantity of data transmitted.

This has implications for policymakers. The enormous value generated by the unlicensed bands was created just by thinking about spectrum policy in a different way. As Lessig pointed out, commons have to be created by the state. For him unlicensed spectrum is a state creation in the same way as are public roads.

And with increasing spectrum scarcity, there are very few bands, garbage or otherwise, that policymakers can make available for innovation at the stroke of the pen. Increasingly, they must be proactive in finding spectrum to de-licence. Regulators are increasingly doing exactly this, and in the United States and Europe there are vociferous discussions about whether this can be done by extending the existing 5 GHz unlicensed band. In this discussion, Wi-Fi advocates fight the satellite and the emerging Intelligent Transport Systems industries.

Yet spectrum alone does not tell the full story of Wi-Fi's success. Although Wi-Fi owes a great deal to fortuitously available unlicensed spectrum, its success probably tells us as much about the value of interoperable standards and well-designed technology as it does about the value of the commons.

Clearly, there are some types of services that do need some sort of licensing regime to avoid interference. The longer the distances involved, the fewer technical solutions are available, and the more one has to rely on political and legal institutions to avoid interference. The current failure of any one standard to transform the IoT in the same way that Wi-Fi-transformed LANs could be seen to be evidence of the limitations of unlicensed spectrum.

For the time being at least, the lessons from Wi-Fi about unlicensed spectrum have been heard loud and clear by policymakers. As Rosenworcel from the FCC said in her Silicon Valley speech: 'we need a game plan for unlicensed spectrum. Unlicensed spectrum can no longer be an afterthought, cobbled together after the fact from junk bands. It deserves attention upfront and in policy primetime'.

11

THE WHITESPACE CONCEPT

'Most of the spectrum, in most of the places, most of the time is completely unused' according to Professor Dennis Roberson of the Illinois Institute of Technology in his introductory comments in Microsoft's video entitled *Dynamic Spectrum Access – Advancing the Digital Economy*. While this might seem at odds with common views of spectrum usage, it refers to the fact that for many radio services, gaps are left between them to try and prevent nearby users causing each other interference. If these gaps or 'whitespaces' could somehow be used, there would be much more spectrum available.

Like ultra wide band, this was an appealing prospect for liberalisers to extract as much value as possible from the airwaves, and in this chapter we focus on early attempts to access the whitespaces in the TV bands.

What Are Whitespaces?

Planning for any radio system requires that spaces are left between transmitters. These spaces are a combination of the following:

- *Geographic gaps* – left to prevent signals from one transmitter causing interference in the service area of a neighbouring one
- *Gaps in the spectrum* – left to take account of the less than perfect performance of filters (see Chapter 3) or to allow transmitters and receivers to operate alongside each other; often these gaps are referred to as guard bands
- *Time separation* – less common but theoretically possible. Imagine a service which operates mostly during the day (e.g. parcel delivery) and another which operates mostly during the night (e.g. taxi service); both could share the same spectrum and interference would rarely occur

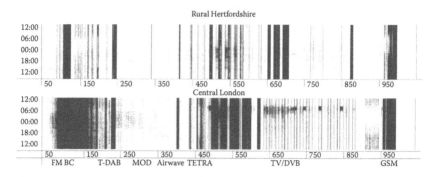

Figure 11.1   The use of 50–1000 MHz at two test sites in the United Kingdom. Note that the darker the colour, the more intense is the usage. White indicates no usage. The full length of the bars shows usage from 12 AM to 12 PM. (From Ofcom, *Spectrum Framework Review*, 2004, pp. 76–77. With permission.)

As can be seen in Figure 11.1 which shows spectrum usage over a 24-hour period at two locations in the United Kingdom, when you combine these factors there are substantial swathes of unoccupied spectrum.

Spaces between transmitters are necessary and the planning rules require them to be implemented to avoid neighbouring services causing each other harmful interference; however, they are generally of most importance for higher powered services where the area over which interference may be caused is potentially large. For many low-power services (such as Wi-Fi), no such planning rules are formally enforced, though good planning can still result in improved performance. Instead, users are free to select any channel or transmitter location that they prefer. More sophisticated users may examine the local radio environment (e.g. which channels their neighbour's Wi-Fi networks are on) and select a channel that minimises interference, but many just select whichever frequency they wish or, where feasible, allow the technology they are using to deal with interference through mechanisms it provides for such a purpose.

As demand for spectrum has increased, a second look has been taken at the spaces left between formally planned services (for which the term *whitespace* was coined) to see whether there could be a means to use them without causing harmful interference. The television broadcasting bands were of particular interest because, as shown in Figure 11.2, the analogue planning rules leave large spaces between transmitters.

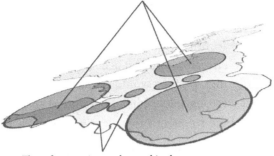

High power TV broadcasts using the same
frequency need to leave spaces between their
coverage areas to avoid interference.

These frequencies can be used in the
"white spaces" in between by lower
power devices.

Figure 11.2   Using the whitespaces in TV transmissions. (From Ofcom presentation: *A Consultation on White Space Device Requirements*, December 2012. With permission.)

In addition, the environment of usage is relatively stable (new transmitters and changes in frequency usage occur over months and years and not hours or days). Other parts of the spectrum have similar whitespaces. For example, it is often said that military spectrum use is most often focused around camps, bases and training grounds rather than over a country as a whole and that it may be feasible to reuse such spectrum in other areas.

Radiomicrophones

Though the concept of the use of these whitespaces seems new, such principles have long since been implemented. For many years, radiomicrophones used the spaces in the VHF and UHF television bands left between the high-power television transmitters. The sharing of the television bands with radiomicrophones works well because of the following:

- The (geographic) gaps between high-power television transmitters can be very large, as the high-power nature of the services and the wide coverage areas requires sites on the same frequency to have a large separation.
- The low-power nature of the radiomicrophones means that any interference that might be caused by them to television

reception would be restricted to a very small area, so much so that any interference caused would likely be within line-of-sight of the use of the microphone such that the impact on television reception would be immediately visible and thus could be dealt with (i.e. by selecting a different frequency for the microphone).

- There is an effective interference self-limiting mechanism in that radiomicrophones would be unable to operate in the presence of a strong television signal and as such they would naturally change frequency to a clear(er) channel.

This latter fact means that in many countries, radiomicrophones operating within television broadcasting bands are licence-exempt and no formal planning takes place. It is up to the user of the microphone to find a suitably clear channel for the location they wish to operate. In some countries (notably the United Kingdom), greater protection of interference to television reception was deemed necessary, and radiomicrophone usage within the television bands is individually licensed. This also serves to provide a means to better manage the spectrum in areas of high radiomicrophone demand (such as the West End theatre district of London) allowing a formal planning process to be put in place.

The model used by radiomicrophones could be seen as a strawman for increased use of the spaces between, in particular, television transmitters. One such method is the use of cognitive radios, which are described in more detail in Chapter 21. Briefly, cognitive radios attempt to follow the same logic used by the radiomicrophone users to identify whitespaces. The idea is simple enough: allow devices to listen to all the channels potentially available to them and identify those on which the weakest signals are present (in the same way as a radiomicrophone engineer might when selecting a channel to use for an event). Two devices that wish to communicate could then compare their lists of 'empty' channels and find one that is clean at both locations and then use this frequency to communicate. In theory this should allow any whitespaces to be identified and permit them to be used for a (lower power) service, filling in the spaces and making more efficient use of the spectrum. If such an approach could be used in the television broadcasting bands (especially UHF bands IV and V), up to 400 MHz of spectrum that has good propagation characteristics for mobile services could be made available.

New Receivers

Many of the planning restrictions for television broadcasting which result in the whitespaces are as a result of the relatively poor performance of early television receivers. For example, the original plans for UHF television in Europe date back to 1961 and require lots of whitespace to be left between and around transmitters to ensure they do not cause harmful interference to their neighbours. With the advent of digital television, many of these planning restrictions no longer apply as the technology used in receivers has developed significantly. Among other things, it has become possible to operate neighbouring transmitters on the same frequency, known as single-frequency networks (SFNs), as well as to operate transmitters on adjacent frequencies without them interfering with each other. Neither of these were possible in the analogue era. Where large spaces had to exist between analogue transmitters, much smaller spaces are needed in an all-digital landscape.

The result of this is that transmitters have become more densely packed, both in terms of the distances between them, and in terms of the spectrum guard bands that were previously required. As a result, the amount of whitespace available has reduced significantly.

To take a specific example, in Europe there were generally 49 television channels available for broadcasting (channels 21–69 inclusive) in the UHF band. In a country with five analogue networks, in any given location, five channels would be in use for delivering a service (e.g. channels 21, 24, 26, 29 and 33 in the example in Figure 11.3). These five channels could not be used in immediately adjacent areas to prevent interference and so adjacent transmitters would use five alternative frequencies. If a seven-frequency repeat pattern were used (i.e. tessellating hexagons as shown in the diagram), there would be six neighbouring transmitters, all using five different channels – an additional 30 channels – which would also be unusable as whitespace as the level of interference from them would be too high. There would therefore be 35 channels that could not be used in each area due to their potential to cause or suffer from harmful interference, leaving 14 (or a total of 112 MHz) potentially free as whitespace (i.e. channels 56–69 in the example above).

Post switch-over, the amount of spectrum available for broadcasting has been reduced as some has been set aside as a new mobile

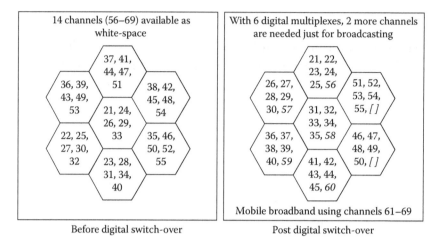

Figure 11.3 Frequency usage before and after digital switch-over.

band (channels 61–69 inclusive – often termed the *digital dividend*), leaving 40 channels available for broadcasting (channels 21 to 60 inclusive). If the same 35 channels were needed for broadcasting, this would leave just 5 for whitespace (only around a third of that originally available). However, as part of the switch-over process, many countries increased the number of terrestrial frequencies used for broadcasting to enable competitive multichannel terrestrial television platforms to develop. Assuming that 6 digital multiplexes were licensed, and using the same 7 frequency repeat pattern, 42 channels would be needed for broadcasting. This theoretically leaves a deficit of two channels required just to deliver broadcasting. The greater efficiency of digital broadcasting in overcoming some of the earlier planning restrictions and the use of SFNs help to reduce the amount of spectrum required; however, it is clear to see that the amount of whitespace available post switch-over is significantly reduced.

Any spaces that do exist in the television broadcasting band can often be reused for more digital broadcasting (e.g. for local television services) due to the improved performance of digital receivers further reducing the opportunity for whitespace.

The same is potentially true for other potential sources of whitespace spectrum. Though a band may be currently full of spaces caused by historical planning requirements, the move to newer, digital technologies may see these bands being able to be used in a more dense and efficient manner, reducing any gaps that might otherwise be available.

Technologies are being developed to take advantage of whitespace spectrum, in the UHF television band in particular:

- 'White-Fi' (officially designated IEEE 802.11af) is a variant of Wi-Fi designed to offer longer communication ranges (e.g. up to 1 km) than standard Wi-Fi (typically no more than 100 m) by using the better propagation characteristics of the UHF band compared to the normal 2.4 or 5 GHz Wi-Fi bands.
- 'Weightless' is a wireless standard for using UHF whitespace that is aiming to provide a means of delivering machine-to-machine communication.

Conclusion

It is worth noting that the term *whitespace* is originally from the printing industry and refers to the space left around an object which allows the object to exist (such as the space between letters). In this context, removing the whitespace would cause the original objects to become impossible to differentiate from each other. There is a strong analogy here for the radio spectrum. Whitespaces may be empty, but they are there for a reason and much care should be taken before filling them in.

Due to, in particular historical, spectrum planning rules, it could be argued that there are many frequencies which are largely unused. Finding a way to use these frequencies (e.g. for a different service that does not need such restrictive planning) could open up a means to access large amounts of spectrum. As these bands become more densely used by their incumbent users, however, the amount of whitespace available may be vastly reduced.

The use of whitespace may therefore be a transient opportunity which only exists until such time as the incumbent use of the spectrum develops to fill in any gaps available.

PART III

THE LIMITS OF LIBERALISATION

12
INTRODUCTION TO PART III

In Part III, we analyse why attempts to liberalise the spectrum market have met with only limited success and consider criticisms of its main policies.

The defining issue here is the lack of confidence in the market to deliver the additional spectrum which mobile operators feel is necessary to meet the growing demand for mobile data services. In most countries, the additional spectrum is being found by the direct intervention of regulators and governments. This suggests that a large-scale reorganisation of spectrum usage can only be achieved by something akin to the traditional command-and-control mechanism. We explore this issue in Chapters 13 and 14.

Where European countries have adopted trading – one of the key tenets of spectrum liberalisation – its use in the high-value mobile bands has been minimal. The main obstacle here seems to be the oligopolistic nature of the mobile markets, an issue explored in Chapter 15. However, we see grounds for optimism in the number of trades in other sectors.

In Chapter 16, we examine whether the economic pressures towards sharing mobile networks are likely to have a negative effect on spectrum liberalisation. The sharing of base station infrastructure is allowed to some extent in mature markets, with operators often pushing to increase the depth of this sharing to reduce their costs. This erodes the separate infrastructures which are the basis for mobile regulation and may endanger competition. The significance of sharing between mobile operators is underlined by the deployment of wholesale networks in the developing world to ensure roll-out to underserved areas or to tackle specific competition issues.

Attempts to liberalise spectrum policy operate within a wider political context and that is the topic covered in Chapter 17. The airwaves emerge as a jealously guarded national asset, and this can frustrate liberalisation initiatives, such as those pursued by the European Commission. The ability to raise funds for the national exchequer by selling off pieces of this precious national asset can also obstruct spectrum policy reform, as we observe in Australia and France. The pursuit of spectrum reform can be problematic where national business and political cultures are less friendly towards liberalisation in general. We discuss this in relation to Thailand, Poland and Iran.

Finally, in Chapter 18 we consider whether auctions have had the unqualified success with which they are usually credited. We examine whether they have perpetrated an oligopoly in the mobile sector, with operators prepared to pay a premium to restrict competition. Are high auction prices to blame for a failure to upgrade mobile networks, as the European Commission has claimed? We also consider whether the trend towards increasing complexity in auctions – particularly the Combinatorial Clock Auction – has damaging side effects.

13
THE CAPACITY CRUNCH

'I believe that the biggest threat to the future of mobile in America is the looming spectrum crisis', Federal Communications Commission (FCC) Chairman Julius Genachowski at a Cellular Telephone Industries Association meeting in July 2009.

When Genachowski made that speech, he was arguably the most powerful spectrum policymaker on the planet. The speech came shortly after WRC-07, in which hundreds of megahertz (MHz) were identified or allocated for the mobile industry. In the Americas, this included the 698–862 MHz band, the 450–470 MHz band and the 2.3–2.4 GHz band. In the United States, the FCC had only just completed its 'hundred-year auction' of the 700 MHz band.

That the speech was made at such a senior level, at such an early stage in the evolution of the mobile broadband revolution, and so shortly after much more spectrum was made available to International Mobile Telecommunications (IMT) reveals just how powerful the spectrum crunch narrative is.

Genachowski, however, is by no means alone in proselytising about the need to allocate and assign more spectrum for mobile technology. For what is probably the majority of stakeholders in the industry, the global effort to overcome the spectrum crunch frames the entire world of spectrum policy. As we are often told, the laws of physics do not allow us to make any more spectrum, but the demand for its use is constantly growing.

One of the first references to a spectrum crunch was in a 2004 paper called *Radio Spectrum Management for a Converging World*, which was produced for the International Telecommunication Union Radiocommunication (ITU-R) sector. Although the paper did mention possible growth in demand for mobile data, it limited its policy discussions to ways of managing spectrum within and among administrations, focussing on methods of spectrum liberalisation. However, the report did not advocate the allocation of swathes of spectrum to IMT.

The pressure to assign more spectrum to the mobile broadband industry grew after the launch of the iPhone in June 2007. From a technological point of view, the iPhone was not a great leap forward. Touchscreen technology was not new and the original iPhone could only access Global System for Mobile (GSM), General Packet Radio Service (GPRS) and EDGE cellular networks (2G, 2.5G and 2.75G), as well as early Wi-Fi technology. However, it was the clean aesthetic and simple user interface, which hugely bolstered the market for mobile data. Consumers increasingly found it more convenient to access the Internet through apps on iPhones (and iPads) than through Web browsers on desktop computers. The huge variety of apps available through Apple products soon became available across a whole ecosystem of smartphones produced by Apple's competitors. Nokia, whose 2G offerings had dominated the market, failed to adapt to the new reality and fell into decline. Such was the growth of demand for mobile broadband that within a couple of years commentators observed that the desktop-based content-neutral browser was becoming upended by walled-garden apps on mobile phones. By August 2010, Chris Anderson and Michael Wolff wrote in *Wired* magazine that 'The Web Is Dead. Long Live the Internet'.

Predicting the Need for More Mobile Spectrum

Probably, the most influential publication on mobile data growth is Cisco's Visual Networking Index (VNI), which has been published annually for almost a decade (Figure 13.1). The VNI predicts how much data will be downloaded around the world over the coming 5 years on a rolling basis. For example, in June 2014, it predicted that 'globally, mobile data traffic will increase 11-fold between 2013 and 2018. Mobile data traffic will grow at a compound annual growth rate of 61 per cent between 2013 and 2018, reaching 15.9 exabytes per month by 2018'. Initially, it was produced internally and was distributed among its partners, but since 2006, its annual updates have become major media events. Its data have been widely cited in spectrum discussions. The VNI does not predict how

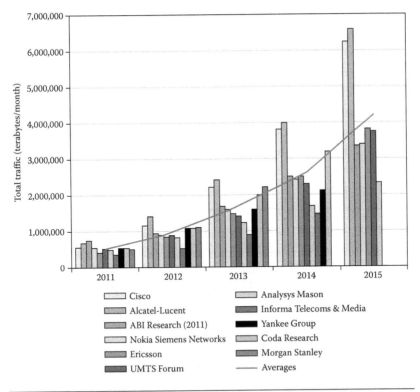

Figure 13.1 ITU-R compilation of mobile data traffic forecasts from 2011. (From ITU-R Report M.2243, 2012. With permission.)

much spectrum will be needed for IMT but the data are sometimes used for that purpose. In June 2014, for example, Robert Pepper, Cisco's vice president for Global Technology Policy attended an annual ITU conference for regulators to argue that the VNI means that global regulators must allow the assignment of both licensed and unlicensed spectrum for IMT.

Before long, policymakers began setting targets for the allocation and assignment of additional spectrum for IMT.

Regulators' Responses

In 2010, The US FCC released its National Broadband Plan (NBP) designed to bring broadband to as many Americans as possible. The plan pointed out that AT&T's mobile network's data traffic had increased 50-fold in the previous 3 years, and cited Cisco's prediction that the demand for mobile data would grow a further 40-fold between 2010 and 2014. In total, the plan called for 500 MHz to be allocated to mobile broadband. As 547 MHz of spectrum was already assigned to IMT, that would create a total of 1047 MHz by 2020.

According to the plan, 300 MHz had to be made available within 5 years – a target which was not met. Measures to find this spectrum include a re-auction of unsold 700 MHz band spectrum, an incentive auction of the 600 MHz band, an auction of the AWS bands (1.9/2.1 GHz) and the release of the 3.5 GHz band. The report said, 'If the US does not address this situation promptly, scarcity of mobile broadband could mean higher prices, poor service quality, an inability for the US to compete internationally, depressed demand and, ultimately, a drag on innovation'.

In justifying the assignment of 300 MHz, the plan says 'the accelerating nature of industry analyst demand forecasts makes clear that it is not a question of *if* the US will require 300 MHz of spectrum for mobile broadband, but *when*.' This prediction, however, was considered 'ultra-conservative' by Cisco Systems.

Indeed, some administrations went much further. In 2010, Sweden's regulator planned to assign an additional 500 MHz (including unlicensed spectrum) by 2015 – 5 years earlier than the United States. China and Russia have also completed their own studies of how much spectrum they need to identify to IMT to avoid a spectrum crunch. The Chinese government has told ITU-R that it needs a total requirement of 1490–1810 MHz by 2020, while the Russian Federation found that it needed a total of 1065 MHz of spectrum for IMT. Towards the end of 2013, Ofcom published its mobile data strategy. The Strategy identified 620 MHz of 'high priority' spectrum that is needed to allow mobile data capacity to increase 47-fold. In a separate submission to the ITU-R, the United Kingdom said its mobile broadband sector could need 775–1080 MHz in a low-demand setting, and 2230–2770 MHz in a high-demand setting.

Intergovernmental institutions also set predictions about the required spectrum allocations to avoid the spectrum crunch.

As outlined in Chapter 17, the European Commission has long had its eye on spectrum, but until 2011 this was largely as an instrument for European harmonisation. However, when preparing the pan-European Radio Spectrum Policy Programme (RSPP), the European Parliament voted to add a new clause to the programme which would require the EU's executive branch, the European Commission to identify 1200 MHz of spectrum for IMT by 2015. The chief advocate of this aspect was the Swedish conservative MEP, Gunnar Hökmark. He said that he hoped the identification of 1200 MHz of spectrum for IMT would mean that 'Europe will be in the forefront of future internet and broadband developments.'*

The RSPP was approved by the European Parliament in February 2012 and the 1200 MHz was to be identified in an inventory report.

The ITU's Role

However, it is only possible to identify more harmonised spectrum for IMT if the ITU-R agrees to a revision to the Radio Regulations, which requires an intergovernmental treaty. The opportunity to do this comes every 3 or 4 years at the World Radio Conference (WRC).

If spectrum is to be allocated to IMT, the ITU-R must first endorse the existence of the spectrum crunch.

The first step down this path was Report M.2078, which was produced in preparation for WRC-07. That report has been enthusiastically interpreted by the telecom industry as an instruction to national regulators to allocate 1300 MHz of spectrum to IMT by 2015.

Report M.2078 was preceded by several reports that determined that an undefined additional amount of spectrum would be required for IMT, and the different methodologies that should be employed to calculate how much.

The ITU-R Report M.2078 (2006) brought much of this research together and concluded that 1280 MHz of spectrum needs to be allocated for mobile broadband by 2020. The report noted that some administrations will not need to allocate this much spectrum, but that some need to allocate as much as 1720 MHz.

As of 2014, these targets had not been met – far less than 1300 MHz has been allocated to IMT in any of the ITU-R's three regions.

---

* *PolicyTracker.* (2011). EU spectrum policy roadmap edges towards final approval. https://www.policytracker.com/headlines/eu-spectrum-policy-roadmap-edges-towards-final-approval/

The ITU-R's next spectrum requirement estimate came 7 years later, in December 2013. The new report (M.2290) estimated that even more spectrum was required by IMT and IMT-Advanced systems by 2020.

For 'low user density settings', the report says 440 MHz is thought to be required for IMT-2000 and 900 MHz is required for IMT-Advanced, a total of 1340 MHz. That is a reduction of 360 MHz for IMT, and an increase of 420 MHz for IMT-Advanced over the 2020 spectrum estimates given in ITU-R Report M.2078. In total, 60 more MHz is thought to be required for mobile broadband in low user density settings than was estimated in M.2078 in 2006.

According to the 2013 report, higher user density settings will require 540 MHz for IMT, and 1420 MHz of IMT-Advanced, which is a total of 1960 MHz. That is a reduction of 340 MHz for IMT, an addition of 580 MHz for IMT-Advanced and a total increase of 240 MHz than what was previously thought. The report emphasised that these figures were a global generalisation and that particular administrations may have lower or higher demand, depending upon their own circumstances.

The report was informed by additional ITU-R reports on the subject, which updated the spectrum calculation methodology, and further information about developments in telecoms. It does not elaborate on which bands should be identified for IMT, only how much spectrum is required. However, the GSM Association, which represents the interests of the mobile industry, has identified four target bands to tackle the spectrum crunch. These are the entire UHF band (470–698 MHz), currently used for broadcasting; the L-band (1300–1518 MHz); the 2.7–2.9 GHz band and the C-band (3.4–4.2 GHz), currently used by satellites.

An Overestimate?

However, these findings are disputed by some industry sectors. While there is little doubt that mobile data demand is growing, particularly broadcasters and the satellite industry question whether the need for extra IMT spectrum is being exaggerated.

The satellite industry was particularly vociferous in their criticism of ITU-R Report M.2290, which informed the WRC-15 process.

Tim Farrar, who does consultancy work for satellite companies, calculated that the ITU-R's estimates of data traffic were as much as 1200 times higher than what the telecoms industry was predicting. To come up with this figure, he benchmarked calculations made from the report's methodology* against forecasts made by the UMTS Forum, a telecoms trade group, and those of Cisco. He says that the UMTS Forum estimates that by 2020 the average amount of data traffic per month per square kilometre across a mean average of urban and suburban areas is 5 terabytes – one million million bytes – whereas the ITU-R predictions suggest that suburban areas will generate 6.4 petabytes – one thousand million million bytes – of traffic per square kilometre per month. From these calculations, he estimates that the model overestimates traffic demand by as much as 1200 times.

However, the GSM Association supports the ITU-R predictions. Their senior director for long-term spectrum, Roberto Ercole, said, 'it is hard to imagine that the ITU tool which has been used and tested for so long (and backed up by independent benchmarking by administrations) can be a factor of more than 1000 out.' He told us that one reason the forecasts are higher than other estimates is that it takes peak demand into account. He said that it is 'peakiness' that drives the need for capacity 'as anyone driving on a motorway on a bank holiday weekend in the UK tends to find out', he said.

The UK input document to Report M.2290, *Study on the Future UK Spectrum Demand for Terrestrial Mobile Broadband Applications*, was criticised by Farrar as an overestimation.

In April 2014, Ofcom quietly updated the webpage that hosted the report to make 'revisions to captions in figures 44, 48, and 49'. In fact, Ofcom had changed the y axis on a graph to change projected growth in mobile data traffic from petabytes to terabytes. In other words, one of the graph's mobile data estimates had been slashed by a factor of 1000. Despite this, the rest of the 100-page report was not changed.

By June 2014, Ofcom was being more circumspect about the report and the need for further spectrum to be allocated to IMT. A spokesperson said, 'our mobile data strategy recognises that

---

* These were set out in a previous ITU document: Report M.1768-1.

long-term demand levels are uncertain and that it is possible, if perhaps unlikely, that beyond 2.3 GHz, 3.4 GHz [planned public sector spectrum release program of Ministry of Defence spectrum] and the proposed release of 700 MHz, there will be little benefit in making more spectrum available'. Nevertheless, the spokesperson also said that it was necessary to secure international agreements so that it could allocate further spectrum, if it wishes to.

Other sceptics scrutinised commercial predictions. For example, in a recent report by Aalok Mehta, and Armand Musey,[*] the two consultants found that Cisco is consistently overestimating the growth in data consumption. They write that over the course of seven forecasts, overestimates were nearly twice as frequent as underestimates (19 versus 10). These overestimates were also on average of greater magnitude than underestimates (103 versus 81 PB/month).

The two consultants argued that the spectrum crunch narrative is widely held because it is promoted by a nexus of in-house research teams within mobile equipment manufacturers that wish to acquire new spectrum, and pliant consultants who are paid to do the same thing. According to them, the lack of resources within modern spectrum regulators allows outlandish predictions to go underscrutinised.

Why Is It Not Hurting?

One argument often made by critics of the spectrum demand projections is a positivistic one – if there is a spectrum crunch, then why are we not noticing it? Most regions and administrations have failed to allocate as much spectrum as they planned to in order to avert the spectrum crunch. Vast swathes of allocated, standardised and sometimes harmonised spectrum – the low hanging fruit of IMT spectrum – are untapped. For example, 4G Americas laments that the majority of administrations in the region had failed to assign even 20 per cent of the 1300 MHz of spectrum defined

---

[*] Mehta, A., and Musey, J. A. (2014). *Overestimating Wireless Demand: Policy and Investment Implications of Upward Bias in Mobile Data Forecasts* (August 15) (http://ssrn.com/abstract=2418364).

in ITU Report M.2078.[*] Similarly, since the adoption of the NBP in 2010, the FCC has managed to assign only 2 × 5 MHz in the H Block, out of the 300 MHz that had to be released by 2015. The 700 MHz band D-block was given over to the critical communications community, and the incentive auction has been repeatedly delayed and is currently expected to take place in early 2016.

Nevertheless, mobile networks continue to function. Therefore, sceptics argue that the spectrum crunch cannot exist or else we would be able to detect it. To them, demands for more spectrum are a cynical ploy to accumulate more spectrum so that operators can cut corners when building networks, bearing in mind the trade-off between the number of cells and the amount of spectrum needed to build a network. Against this, one could equally argue that we can already detect the oncoming spectrum costs in the high prices that consumers have to pay in some countries for their mobile broadband.

Sceptics also point to the frequent failure of spectrum auctions. While clearly some fail because they have been badly designed (Australia), sometimes regulators simply cannot persuade operators to take the spectrum off their hands (Cyprus). Therefore, they argue that there is already too much spectrum available for mobile broadband.

This scepticism seems to be permeating through to some policy-makers, as can be seen in the RSPP's inventory report, which was published in September 2014. The report noted that there already was 1000 MHz of harmonised spectrum available to IMT, but that the majority of member states had not assigned all of it, particularly in the 3.4 GHz band.

Regarding the spectrum crunch, the inventory report was circumspect. It read, 'Based on the analysis of spectrum supply and demand, the Commission believes there is currently no need for additional spectrum harmonisation, beyond the 1200 MHz target, in the range 400 MHz–6 GHz for licensed wireless broadband'.

What nobody is doubting is that there is a massive growth in demand for mobile data, and that more spectrum will be required for

---

[*] *PolicyTracker.* (2013). Latin American mobile industry players attack 4G policy errors. https://www.policytracker.com/headlines/key-latin-american-mobile-industry-players-blame-defective-government-policy-for-laggard-4g-roll-out

mobile broadband, especially if mobile broadband comes to be seen as a utility as vital as water or electricity. As to what extent the nightmarish scenarios of a spectrum crunch will occur, only time will tell.

However, a central paradox remains: liberalisers believe that spectrum should be allocated by the market and this policy has been enacted in many of the areas mentioned: the United States, the European Union and the United Kingdom. So why can we not leave it to the market to find the new spectrum needed for IMT?

14

WHY IS LIBERALISATION LIMITED IN ITS APPLICATION?

The commercial value of mobile networks inspired spectrum liberalisation, but their spectacular success also exposed the limitations of the project. As explained in the previous chapter, the mobile industry entered a new phase of growth in 2007 when the smartphone market really started to take off with the launch of the iPhone. By 2009, it became clear that mobile phone users globally were using increasing amounts of data.

The increase was steep and the exponential growth predicted by companies like Cisco and Yankee led the mobile industry to believe it would soon run out of spectrum. This argument was widely accepted, and in mid-2010, the United States was the first to announce its regulatory response,[*] promising to make 500 MHz of additional spectrum available for wireless broadband.

Within a year Sweden,[†] the United Kingdom[‡] and Denmark[§] had made similar announcements and the policy of finding extra spectrum for wireless broadband has since been adopted practically worldwide.

But the key point here is what the mobile operators *did not* do. All the four countries above had market-friendly regimes – why did no

---

[*] *PolicyTracker.* (2010). US government plans to double available wireless spectrum. https://www.policytracker.com/headlines/us-government-plans-to-double-available-wireless-spectrum

[†] *PolicyTracker.* (2010). Sweden to use 800 MHz auction to achieve universal broadband coverage. https://www.policytracker.com/headlines/sweden-to-use-800-mhz-auction-to-ensure-universal-broadband-access

[‡] *PolicyTracker.* (2010). UK to find 500 MHz of spectrum for "superfast" broadband. https://www.policytracker.com/headlines/uk-to-find-500-mhz-of-extra-spectrum-for-201csuperfast201d-broadband

[§] *PolicyTracker.* (2011). Danes seek 600 MHz of extra spectrum for wireless broadband. https://www.policytracker.com/headlines/danes-look-for-600-mhz-of-extra-spectrum-for-wireless-broadband

mobile operator seek to buy spectrum from lower value users, like broadcasters? Why did no regulator remove the restrictions on the use of TV spectrum and let the mobile industry negotiate directly with the broadcasters?

No Faith in Our Creation

One simple answer is that nobody expected the market to deliver the results within the required time scale. The mobile industry's response was to lobby governments and regulators for a larger spectrum allocation – effectively a return to the old command-and-control system and a policy approach, which has been accepted as the solution to this problem by governments and regulators worldwide.

If a farmer with new machinery and the latest fertilisers could get much higher land yields than his or her neighbour, why would the farmer not buy, or rent, the fields next door? If this does not happen, there must be obstacles blocking the market.

The 'data tsunami' episode exposed the existence of these blockages. These have been confirmed in 2012 to 2014 and it has become plain that TV will be moved out of 700 MHz in Europe, not by trades or commercial negotiation, but by the actions of regulators and governments and ultimately in the time-honoured manner – at the International Telecommunications Union (ITU).

Obstacles for the Development of Spectrum Markets

Time Lags

What is obstructing the development of a market in spectrum?

The first problem is its slow-moving nature. This is caused by the economies of scale and long product development cycles required in the mobile industry but also in other sectors which depend on spectrum.

Transmitters make up a significant proportion of the cost of a mobile handset and in order to keep the retail price down to an affordable level, manufacturers have a target price of US$1 per transmitter chip. Chip prices only come down when they are produced in the millions, and the industry rule of thumb is that chips must address markets of 200 million people and over. Only China, India, the United States, Indonesia and Brazil are big enough on their own, but even so, a regional market is a safer bet than a national one.

For example, a handset manufacturer is about to spend huge sums on developing a handset, which works in XXX MHz. This process takes at least 3 years, irrespective of the spectrum issues.

How does the manufacturer ensure that XXX MHz will actually be a mobile band in a wide enough target market in 3 years' time?

The manufacturer needs buy-in from a wide range of countries and this is what the current ITU system delivers. The band could be proposed for discussion at the next World Radio Congress in 4 or 5 years' time, regional groupings could support this and in the best possible scenario there could be a near-global identification of XXX MHz as a global band.

Back in 2009, when an imminent spectrum shortage seemed likely, if mobile operators had tried to buy up spectrum in the XXX MHz band, there was a danger of fragmentation. Broadcasters in country A with very little terrestrial television may be prepared to sell their spectrum because most viewers watch by satellite or cable. But the reverse may be true in neighbouring country B, where 50 per cent of viewers use terrestrial TV.

The mobile operator may succeed in buying spectrum in one country, but not in many others, leaving it with a target market way below the necessary 200 million.

At One Remove

A further problem for liberalisation is the separation of mobile operators and manufacturers. It is mobile manufacturers who produce handsets but it is mobile operators who buy spectrum.[*] This loads up the risk for any mobile operator who makes preemptive purchases without knowing for certain that handsets would be available.

---

[*] The capital arm of the chip-maker Intel tried to get around this problem by buying spectrum in the Swedish 2.6 GHz auction in 2009, hoping that this would be passed on to operators using its favoured WiMAX technology. However, this is very unusual. WiMAX was not successful and Intel later sold the licences.

In the XXX MHz example above, an operator would be spending hundreds of millions of dollars but needing to rely on another entity to provide the handsets. They could certainly enter into a commercial agreement with a handset manufacturer, but there is always the possibility of the other company going bankrupt or pulling out of the deal, ratcheting up the risks when the operator looks for outside financing.

The Global Nature of the Spectrum Market

Taking a step back from the spectrum world for a moment, it is often pointed out that we live in an increasingly globalised world, and this has made markets more competitive. Manufacturing in the developed world has been hit by lower labour costs in newly opened markets in developing and mid-income countries such as China. Workers in knowledge-based industries like computer programming can now face competition from around the world, particularly software power houses like India.

But for spectrum users, the global nature of the market restricts competition rather than encourages it. We have just talked about one aspect of this: the need for chip manufacturers to target markets of 200 million and above, making the market for equipment at least regional or preferably global.

There is another aspect to the global nature of the spectrum market. Transmissions, as we are often told, do not recognise national boundaries and can easily cause interference in a neighbouring country. Adjacent countries therefore need to coordinate their spectrum use, and this causes a chain effect whereby France's use of frequency XXX for TV could prevent it being used for mobile in Ukraine, over 1000 km away. Interference would prevent France's neighbour Germany using the frequency for mobile at its borders, and hence, it would tend to use it for TV, having the same effect on its neighbour Poland which borders Ukraine (Figure 14.1).

Hence, the main frameworks for frequency use are worked out in regional technical organisations, such as the European spectrum regulators' organisation, the European Conference of Postal and Telecommunications Administrations (CEPT), and these are coordinated via a global group, the ITU. The physical nature of spectrum

Figure 14.1 Central Europe. (From Generic Mapping Tools and ETOPO2. With permission.)

creates a bias towards international homogeneity, which favours bigger players and tends to militate against the 'many buyers/many sellers' model of a perfectly functioning market. The physical nature of land – the key business input in the agricultural sector – does not have the same effect. How one farmer uses the land (i.e. what crops he grows) does not necessarily prevent his neighbour planting something different and has no effect on what farmers in other countries are able to grow.

A further aspect of globalisation in this field is the international nature of some businesses, which use spectrum. Satellite uplinks could be based in one of many countries, so operators can shop around for the regulatory environment which suits them best. This means any liberalising country is at a disadvantage if it applies spectrum charging when other nations do not. Satellite operators could simply move their uplinks to a cheaper jurisdiction.

There is a similar problem with maritime and aeronautical spectrum, as discussed in Chapter 6. Ports and airports in liberalised countries would have to raise their charges to shipping or aircraft to cover new spectrum pricing measures, putting them at a disadvantage compared with international competitors.

So unlike many other sectors, the global nature of spectrum puts a break on market mechanisms, rather than creating new opportunities.

The Political Power of Broadcasters

Television is the most popular news medium in most countries,[*] giving it a huge political power. Politicians need a medium to communicate with the voters: TV is the most popular medium and also the most trusted. For mobile operators, getting access to TV spectrum would be like prising a limpet off a rock.

Despite the rise of the Internet and mobile telephony, TV viewing remains remarkably stable or is even increasing according to some studies.[†] It averages around 4 hours a day in the United Kingdom and 5 hours a day in the United States. Ed Richards, chief executive officer (CEO) of the UK regulator Ofcom, described TV as 'the cockroach of the internet apocalypse…whatever the internet revolution throws at it, TV survives and survives and survives'.

Television is often said to 'punch above its weight' in the world of spectrum lobbying. The mobile industry sometimes complains about this, arguing that it makes a much greater contribution to national economies.

However, we should not view television's power in too cynical a light. It is not purely a matter of mutual back-scratching between politicians and the TV industry. TV is the predominant means for passing down national cultures through the generations, the primary source of entertainment for most of us, the first port of call for citizenship information, a vital tool for integrating the ethnic and geographic diversity of the modern nation state and even a source of companionship for the old.

In short, the TV industry has great lobbying power not just because it has a symbiotic relationship with politicians but also because it is

---

[*] To take two examples, see Olmstead, K., Jurkowitz, M., Mitchell, A., and Enda, J. (2013). How Americans Get TV News at Home. Pew Research Center (http://www.journalism.org/2013/10/11/how-americans-get-tv-news-at-home/) for the United States and Gisby, J. Industry Perspectives on the Future of Commercial Communications on TV and TV-Like Services. An independent report commissioned by the Office of Communications (Ofcom) (http://stakeholders.ofcom.org.uk/binaries/research/tv-research/Future_of_Commercial_Comms.pdf) for the United Kingdom.

[†] See TV Viewing Figures Increase in UK. BBC News, Entertainment & Arts (http://www.bbc.co.uk/news/entertainment-arts-21828961) and On Average Viewers Watched 3 Hours 52 Minutes of TV per Day in 2013, Down by 9 Minutes from 2012. Ofcom (http://stakeholders.ofcom.org.uk/market-data-research/market-data/communications-market-reports/cmr14/tv-audio-visual/).

a vital part of most people's lives. One can question the efficiency of broadcast delivery platforms, but it is much harder to question the usefulness of the medium.

The importance of TV means it is enshrined in law and policy in most countries. Many countries have public services broadcasters (PSBs) who have public policy goals such as providing news and citizenship information; both PSBs and commercial broadcasters are typically required to make their programming available to the whole population. Many broadcasters use spectrum to meet this universal service requirement, meaning that they would be breaching their legal obligations if they did sell it to mobile operators.[*]

Even if it were possible to meet universal service requirements via cable or satellite, in most countries this would be a major policy change which could not happen without political approval. Furthermore, TV is generally regarded as so important that democratic oversight is seen as entirely appropriate.

So the option of buying spectrum from broadcasters must have been such an impractical solution to the capacity crunch that mobile operators did not even consider it. It is highly unlikely that broadcasters would have the will or ability to sell spectrum because they are so enmeshed in an existing political structure, which imposes specific obligations and constrains their behaviour.

It is far more practical to lobby ministers and regulators for more spectrum, as any revision of broadcasting policy is likely to require government intervention. This has the further advantage of combining neatly with the traditional ITU route to spectrum allocation, which also offers access to the large markets needed to create economies of scale.

A Wider View

The bigger issue is whether there is any hope of spectrum moving to higher value uses or better technologies through the existence of a more open market.

The experience of the past decade is that liberalisation is allowing operators to move to superior technologies without any need for regulatory intervention, but change of use is rare.

---

[*] See Chapter 7 for a more detailed discussion.

One example of technological progression is the move from WiMAX to Long Term Evolution (LTE) in 3.5 GHz, which started in about 2012 and was made possible by technology-neutral licencing.[*] Another is the 2009 repeal of the GSM Directive,[†] which made it possible to use 900 and 1800 MHz for services other than GSM. Within a couple of years, this led to the widespread launch of LTE data services in 1800 MHz and plans for the use of 900 MHz for LTE voice services.

Attempts to change the sort of service supplied in a band seem to be beset by problems similar to those preventing access to broadcasting spectrum. Nobody has seriously suggested that the military should compete in an open market to buy the spectrum it needs. This is because of understandable government interest in national security and also the need to coordinate with other powers in organisations such as the North Atlantic Treaty Organization. Even attempts to create a quasi-market, such as that recommended by the Cave Review in the United Kingdom in 2005,[‡] have proved extremely difficult to implement, and it will be 11 years before the first military spectrum is opened up to the private sector.[§]

Where military spectrum has been sold to the private sector in other countries, this has usually been achieved through an administrative re-farming led by the regulator and arguably this is what the UK process ultimately amounts to.

There is perhaps more hope for liberalisation in the story of Qualcomm's MediaFLO technology, which was designed to deliver mobile TV. Mobile TV did not take off in the way pundits had predicted, and the technology was problematic in that it required the building of a separate network.

---

[*] https://www.policytracker.com/headlines/low-key-uk-broadband-switches-on-uk2019s-first-lte-network

[†] https://www.policytracker.com/headlines/eu-legalises-refarming-of-2g-spectrum

[‡] *PolicyTracker*. (2005). Spectrum Audit makes proposals to shift public sector thinking. https://www.policytracker.com/headlines/spectrum-audit-makes-proposals-to-shift-public-sector-thinking

[§] 190 MHz of spectrum in the 2.3 and 3.4 GHz bands is due to be sold in 2016 (*PolicyTracker*. (2013). Ofcom given responsibility for military spectrum auction. https://www.policytracker.com/headlines/uks-ministry-of-defence-hands-over-a-tranche-of-spectrum-to-ofcom).

In the United States, Qualcomm bought 700 MHz spectrum in order to build this network but was able to sell this at a profit to AT&T in 2010 when they decided to pull out of the market.* AT&T planned to use this for LTE, a move which sounds very close to the liberalisers' dream of spectrum flowing to superior technologies and higher value uses.

But was this really a change of use? That is a matter of interpretation. The focus of MediaFLO was the delivery of live and prerecorded TV and audio to mobile handsets. The LTE also delivers video to mobile handsets, and it is superior to MediaFLO in that it provides the full range of broadband data services as well. It is inferior in the sense that its transmission of live programmes is less spectrally efficient than MediaFLO.

It could be argued that the use here is consistent – video services for mobile phones – but LTE is more commercially viable because it offers other valuable services without the need to build a separate network. Is this a change of use or a change of technology? The weight of argument leans towards the latter – MediaFLO was only ever intended for the delivery of video to mobiles – it would not have worked for big-screen TVs.

So while Qualcomm's spectrum sale was certainly good news for the liberalisation camp, it is hard to find a clear case in recent years of market mechanisms allowing spectrum to migrate to a higher value use. The best example remains Nextel in the United States, which created a public mobile network out of spectrum designed for a business radio despatch network. This happened in the early 1990s, and while it inspired a generation of liberalisers, the policies they pursued have not created a new wave of Nextels.

---

* *PolicyTracker.* (2010). Qualcomm makes impressive profit from spectrum sale. https://www.policytracker.com/headlines/qualcomm-makes-impressive-profit-from-700-mhz-spectrum-sale

15

WHY DOES TRADING HAVE SUCH PATCHY SUCCESS?

Allowing the trading of licences has been a key part of the liberalisation agenda. It has been in operation in the United States and Australia since the 1990s, and was introduced in Europe's most enthusiastic spectrum liberaliser, the United Kingdom, in 2004. France and Germany quickly followed suit and the Radio Spectrum Policy Programme (RSPP)[*] agreed by the European Parliament in 2012, requires all EU member states to allow spectrum trading in the main mobile bands by 2015.[†]

The economic theory behind spectrum trading is that it should create a more open market by making it quicker and easier to get access to the airwaves. If spectrum can only be licensed by a regulator or government, then it is usually illegal to sell it to someone else directly or the only way would be to sell the whole company. This has what economists call high transaction costs – the outlay associated with making this sale may be so great as to outweigh any benefits. Transferring the spectrum assets into a new spectrum-only company which could be sold would involve expensive legal costs and a time-consuming administrative burden.

It is far better to allow licences to be traded directly between licence owners with regulators or governments only becoming involved if there are technical or competition issues. The simpler the process, the easier it is for existing spectrum users to move to other bands or adopt a wired alternative if a higher value user makes an offer for their frequencies.

---

[*] European Commission. Digital Agenda for Europe. Improving Access to Radio Spectrum through Market Mechanisms (http://ec.europa.eu/digital-agenda/en/improving-access-radio-spectrum-through-market-mechanisms).

[†] See *PolicyTracker*. (2012). EU lawmakers approve RSPP. https://www.policytracker.com/headlines/european-parliament-approves-rspp. The bands are 790–862 MHz, 890–915 MHz, 925–960 MHz, 1710–1785 MHz, 1805–1880 MHz, 1900–1980 MHz, 2010–2025 MHz, 2110–2170 MHz, 2.5–2.69 GHz and 3.4–3.8 GHz. These are also harmonised throughout the European Union.

Poor Uptake

The problem for liberalisers is the lack of spectrum trades in Europe, particularly in the mobile sector.

This contrasts sharply with Australia, where the Productivity Commission estimates the annual turnover of spectrum licences at 6 per cent, similar to the commercial property market.[*] In the United States, deals worth hundreds of millions of dollars are not unusual. At the beginning of 2014, T-Mobile announced a mutual trade with Verizon Wireless estimated to be worth $3.3 billion. Meanwhile US telecoms giant AT&T said it would acquire 49 AWS licences covering 50 million people from Aloha Services.

The United Kingdom provides the best information on spectrum trading activity, but even here giving exact figures is difficult because the Transfer Notification Register[†] shows both the sale of whole companies and 'pure' spectrum trades where the company is only selling its licences but continues to trade. For example, in the construction industry Lafarge Aggregates Limited bought Hope Construction Materials Ltd. outright in 2014, and the transfer of the business radio licences was recorded in the register.[‡]

So assessing exactly how many were pure trades and how many were company sales is difficult, but we can be certain of one thing: Europe has not followed the US pattern of multimillion pound trades of mobile licences. The main mobile bands have been tradable in the United Kingdom since 2011, but no voluntary trades have taken place. Of the three transfers recorded,[§] two reflect the merger of Orange and T-Mobile to create EE and one shows the

---

[*] *PolicyTracker.* (2008). Architect of UK spectrum liberalisation 'disappointed' by number of trades. https://www.policytracker.com/headlines/architect-of-uk-spectrum-liberalisation-disappointed-by-number-of-trades

[†] See Ofcom's *Spectrum Trading Register* available online at: http://spectruminfo.ofcom.org.uk/spectrumInfo/trades

[‡] See Data available from Ofcom's *Spectrum Trading Register* at: http://spectruminfo.ofcom.org.uk/spectrumInfo/trades?groupTradeRef=TNR-2014-09-014 and information about the company available on Wikipedia here: http://en.wikipedia.org/wiki/Hope_Construction_Materials

[§] Data available from Ofcom's *Spectrum Trading Register* at: http://spectruminfo.ofcom.org.uk/spectrumInfo/trades?after=dd%2Fmm%2Fyyyy&before=dd%2Fmm%2Fyyyy&seller=&buyer=&service=Public+Wireless+Networks&submit=Filter+results

sale of spectrum from EE to Three, which was required by the regulator to alleviate competition concerns created by the sale of 800 MHz and 2.6 GHz.

We are not aware of any trades of mobile spectrum anywhere else in Europe, so it is disappointing for liberalisers that there is so little activity in what is the highest value sector.

Concentration and National Licences

One reason for this is the oligopolistic nature of the mobile market. 'Perfect markets' have a large number of buyers and sellers, whereas most European countries have only three or four mobile operators. There are also barriers to entry: the mobile industry requires such high capital investment that there is no longer a host of new entrants snapping at the heels of the existing players. It is more a question of incumbents competing with incumbents, and to sell spectrum in that pressure cooker environment would be a sign of weakness. Investors and the stock market would interpret any sale of the operator's defining asset as a signal of imminent market exit, leading to a loss of confidence in the company and its shares.

But why is the situation so different in the United States? In Europe, mobile licences are almost exclusively national, but in the United States they are regional: in 2014 the 1900 MHz auction was divided into 174 separate licences. This jigsaw approach means operators often want to fill coverage gaps, sowing the seeds of a more dynamic market.

The Verizon/T-Mobile trade mentioned earlier was in fact a mutual exchange of spectrum, but with the partly German-owned company getting the lion's share and paying an additional $2.4 billion. This was not seen as a sign of weakness by the stock market as there were advantages on both sides: T-Mobile was in fact paying $1.85/MHz/pop for spectrum, which Verizon had bought for $1.47/MHz/pop in 2008.[*]

Deals which appear mutually beneficial are much harder when national licences are at stake: even though the seller may generate cash, the overall signal is negative, suggesting a reduction in its customer offer or a reluctance to plan for the future.

---

[*] *PolicyTracker.* (2014). T-Mobile and Verizon agree $3.3 billion spectrum trade. https://www.policytracker.com/headlines/t-mobile-and-verizon-in-3.3-bn-spectrum-trade

Technologies in Decline

The disappointing number of high-value trades in Europe may also be explained by the stage in the technology cycle. Many of the big ticket sales in the US spectrum market concerned the reallocation of frequencies intended for mobile TV. This applies to both the Qualcomm sale of 700 MHz spectrum designed to be used for MediaFLO (See Chapter 14) sale and Aloha's 2008 and 2014 trades of spectrum in the same band, which it had planned to use for DVB-H services.

There was also great interest in buying the 2.6 GHz licences used by Clearwire for WiMAX services, but in 2013 the company was acquired by its majority shareholder, Sprint, despite intense competition from the DISH network. This meant that the licences came into new ownership by company acquisition rather than trading.

The pattern here is that both mobile TV and WiMAX were technologies which failed to take off, meaning that operators using Long Term Evolution were keen to take advantage of the unused spectrum. In Europe very few licences suitable for mobile TV alone were offered, so they could not become available in the secondary market. From 1995 through 2010, the United States had auctioned more spectrum than Europe and allowed trading much earlier, so there was a larger pool to fish.

The sheer size and wealth of the United States arguably encourages entrepreneurial activity, and if these risky ventures fail the spectrum can be resold in the secondary market. With a potential market of 313 million people in the world's richest country, the economics of a new mobile TV network in the United States must have seemed like a reasonable risk. However, the average population of an EU member state is about 18 million, and some of these are far from wealthy, blunting the entrepreneurial impulse and producing less spectrum on the secondary market.

Achievements of Spectrum Trading

There have been a considerable number of trades in the lower value sectors. The UK regulator Ofcom has identified the fixed links market as a particular success,* and there have been 90 licences in this sector that have been sold or traded in the past 10 years.

As explained above, some of these reflect purchases of whole companies, but many of them do not. Considering the 22 fixed links trades since the start of 2014, there are 12 by UK Broadband, most involve multiple licences, and the buyer in most cases is a local council.† These are 'pure' spectrum trades: UK Broadband continues in business providing Internet connectivity services for business and consumers.‡ There are also three trades involving multiple licences where a new company, Tixos, has bought about 200 links from two other firms,§ without buying the whole organisation.

So there have been at least 15 pure spectrum trades in the fixed links sector in 2014, which seems a reasonable success, certainly when compared to the lack of activity in the mobile market.

Liberalisers argued that spectrum trading would make it easier for vacant spectrum to be reused and for frequencies to naturally migrate to higher value users. In practice, this facility is reasonably well used *if the circumstances are right.*

Fixed links trading seems to be popular in the United Kingdom because it does not have the anti-competitive characteristics of the mobile market. There are many buyers and sellers: scores of independent contractors provide fixed links, and some companies in the energy sector set them up in-house. The barriers to entry in the fixed links market are not particularly large, nothing like the billions required to set up a mobile network.

Trading in the lower value licences has also been quite successful in France with a reasonable number of trades in fixed broadband wireless access licences in 3.4–3.6 GHz. Here the authorisations

---

* *PolicyTracker.* (2008). Architect of UK spectrum liberalisation 'disappointed' by number of trades. https://www.policytracker.com/headlines/architect-of-uk-spectrum-liberalisation-disappointed-by-number-of-trades

† Data available from Ofcom's *Spectrum Trading Register* at: http://spectruminfo.ofcom. org.uk/spectrumInfo/trades?service=Fixed+Links&submit=Filter+results&before= dd%2fmm%2fyyyy&after=dd%2fmm%2fyyyy&page=1

‡ See the company website here: http://www.ukbroadband.com/

§ See the following entries from Ofcom's *Spectrum Trading Register*: http://spectruminfo. ofcom.org.uk/spectrumInfo/trades?groupTradeRef=TNR-2014-06-001 and http:// spectruminfo.ofcom.org.uk/spectrumInfo/trades?groupTradeRef=TNR-2014-08-014

were granted on a regional basis, and spectrum trading has made it possible for operators seeking national reach to consolidate their licences.[*]

In conclusion, spectrum trading has not been a failure. It seems to be creating a useful benefit in the lower value sectors of the market. But there are many obstacles to spectrum liberalisation and while trading has helped reduce one of them – high transaction costs – the effect has been marginal rather than revolutionary.

---

[*] See UK 'celebrates' 10 years of spectrum trading. (2014). *PolicyTracker* 11 December (https://www.policytracker.com/headlines/spectrum-trading-10th-anniversary).

16

MOBILE NETWORKS

From Sharing to Wholesale*

When the mobile industry took off in the 1990s, it was a breath of fresh air for economists because it offered infrastructure competition that overcame the restrictive single pipes of national fixedline telecoms. There was also competition within the industry. There were several mobile operators, each with their own network, each competing to use them more effectively and so attract more customers. However, as the new millennium progressed, there was a trend towards the sharing of networks to reduce costs and improve coverage. And from 2010 onwards, with the advent of Long Term Evolution (LTE), there has even been increasing discussion of wholesale mobile networks that are built through public–private partnerships (in the same vein as governments investing in wholesale national fibre roll-out). The idea of governments commissioning their own networks was unthinkable a decade ago, but these are expected to be deployed in a handful of countries from 2015 onwards.

The more elements of a network that are shared, the greater is the impact on spectrum liberalisation. If there is no sharing, operators' use of separate networks creates the possibility of a spectrum market but at the other end of the scale, if all operators were to run a joint network, they would only be able to compete on services. In effect, they would be buying network capacity from a shared network, rather than buying or selling spectrum.

The challenge for spectrum liberalisation is that the simple 'competing networks' model is being undermined in both mature markets and developing countries. Many richer countries allow network sharing between operators and in less well-off nations wholesale networks are gaining acceptance as a way of reaching improving coverage in underserved areas or as a response to specific regulatory difficulties.

---

* This chapter was written with the help of Catherine Viola.

Spectrum liberalisation has survived this challenge so far, but if wholesale mobile networks became the norm, current thinking about market mechanisms would require a major overhaul. The use of the airwaves by the mobile sector could become more like the regulated monopoly of the fixedline market.

In this chapter we consider the industry pressures which have led to network sharing, the impact of LTE and the arguments behind moving to a wholesale model.

Why Share Mobile Infrastructure?

By sharing their infrastructure, operators can achieve a series of benefits: cost-efficient capital investments, slicker operations and reduced operational costs, a better network and user experience, quicker roll-out of new services, faster market entry, enhanced revenue-generating opportunities and improved customer retention. Taken together, the commercial benefits arising from sharing enable an operator to improve its competitiveness and, ultimately, its profitability. Infrastructure sharing can therefore be considered as a strategic tool for increasing shareholder value.

For years, those operating in mature, saturated mobile markets have been battling static or falling average revenues per user (ARPUs) and squeezed margins. At the same time, operators need to fund capital investments to upgrade their networks and increase capacity to cater for growth in data traffic. New licences often come with stringent rural coverage obligations, bringing network sharing and spectrum pooling to the fore as cost-effective means of providing sufficient coverage and data rates.

In these circumstances, operators are under pressure from their shareholders not only to reduce their operating costs but also to make the most efficient use of their capital investments and to effectively manage risk (Figure 16.1).

A distinction is usually made between active and passive sharing. Passive sharing is commonplace and involves the sharing of the physical base station infrastructure, such as the compound but frequently also mast/towers and the equipment room. This lowers site rental costs, reduces the total number of sites needed and also accelerates roll-out.

Active infrastructure sharing is more complex but involves some sharing of the electronics in a base station, frequently including the

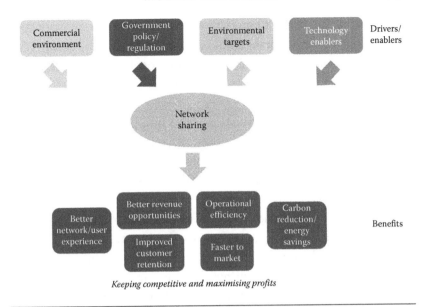

Figure 16.1 Drivers, enablers and benefits of mobile infrastructure sharing.

radio access network (RAN). This poses more difficult regulatory issues as well as strategic issues for the sharing partners, and has therefore been slower to take off.

Potential Savings

Infrastructure sharing in its various forms, from passive site sharing through to full RAN sharing, offers scope for operators to make significant cost savings over the longer term. The deeper the level of sharing, the greater the extent of the potential cost savings, and there are benefits for both capital and operational expenditure:

- The capital costs of building new sites are halved because they are shared between the two parties (most arrangements are bilateral).
- In RAN sharing arrangements, sourcing network equipment and services as a combined entity helps increase economies of scale and reduce costs.
- Site consolidation results in fewer sites overall, bringing opex savings for both parties.
- Operational efficiency can be further improved by combining the parties' operations and maintenance functions.

Operators can expect to save 20–40 per cent on their network costs (depending on the implementation) by entering into bilateral RAN sharing arrangements. Even greater savings could be achieved if sharing were extended to three or more parties – subject to regulatory approval.

In very broad terms, regulation is becoming more supportive of infrastructure sharing. In many countries, passive infrastructure sharing is actively encouraged or even mandated by regulation to reduce the proliferation of mobile masts, with the result that site sharing is now widely practised worldwide.

Attitudes are changing, too, towards RAN sharing. Regulators remain cautious in their approach to sharing active network elements, backhaul and spectrum, due to concerns about restricting competition, service differentiation and consumer choice. Nonetheless, evidence suggests that policymakers and regulators are more open to considering active infrastructure sharing than in the past.

The environmental benefits of infrastructure sharing are another driver. Energy (and cost) savings can be achieved by sharing masts, antennas, power supplies and air-conditioning. At the same time, sharing also reduces the visual impact of mobile networks, by stemming the proliferation of masts.

A further pressure towards active network sharing is the ready availability of technological solutions. For some years, a range of operators have been marketing solutions for 2G and 3G which are a great improvement on the workaround network and handset solutions used for the earliest RAN sharing implementations.

LTE: A Boost for Network Sharing

However, with LTE there is a major step change. LTE's RAN sharing features are much more sophisticated, greatly increasing its attractiveness as a commercial proposition. From 3GPP release 11 in 2012, sharing operators can distinguish between their respective services and there is no need for proprietary mobile terminals to access the shared services. Up to six sharing operators can be accommodated.

These features will be enhanced in 3GPP release 13,[*] in 2016, which is due to the following:

- Asymmetrical sharing of RAN resources or dynamically modifying their allocation to meet criteria such as the financial interests of the sharing parties
- Load balancing between cells based on the financial contributions of the sharing network operators
- Variable sharing of network capacity at different times of day

In simple terms, LTE has already made huge strides to make RAN sharing easier technically, and it should soon allow single networks to be managed in different ways to meet the needs of different operators. In terms of resource allocation, customer information, billing and service differentiation, the technical advantages of having your own network are being eroded.

A further technical pressure towards shared networks is LTE's need to operate in wide bandwidths to achieve the highest data rates. These bandwidths should ideally be 20 MHz (paired) or more; in bandwidths of 5 or 10 MHz, there is little improvement in performance over 3G HSPA+. Particularly with the 800 MHz band (which is important for wide area coverage), bandwidth is limited, resulting in either just a few licences with wide bandwidths or a greater supply of licences with lower bandwidths.

Some operators are also struggling to find a viable business model for deploying LTE. A wholesale network provision model would overcome both these difficulties and allow LTE to run in optimal bandwidths as well as lower network costs to support the LTE business case.

Wholesale Networks

The model for wholesale network sharing could be a commercially negotiated multiparty collaboration, where two or more operators divest their network assets to a third-party network company (netco) and then buy capacity on demand. Service-level agreements and key

---

[*] See 3GPP report TR 22.852, *Study on Radio Access Network (RAN) Sharing Enhancements* (http://www.3gpp.org/DynaReport/22852.htm).

performance indicators would govern the service provided and allow for differentiation between the sharing operators' services.

Another model would be a single open-access network available to all market participants, including operators, Mobile Virtual Network Operator (MVNO) and resellers. This network might be limited to rural areas or could be nationwide. Implementing this type of model is likely to be easier with a greenfield network roll-out such as LTE.

There is significant interest in an open-access model for LTE in Africa. In Kenya, the government is inclined to overrule the wishes of incumbents and proceed with developing a single open-access network for LTE, rather than auctioning limited frequencies in the 800 MHz band (see Box 16.1).

BOX 16.1   AN OPEN-ACCESS LTE NETWORK FOR KENYA

At the time of writing Kenya was moving forward with plans to deploy an open-access network for LTE. Driven by the scarcity of 800 MHz spectrum (reportedly only 25 MHz is available), the government decided to create a public–private partnership, known as the '4G Special Purpose Vehicle', to own and finance the network deployment.

Taking part in the venture are the country's mobile operators, three broadband providers and three equipment vendors. These partners will acquire equity in the network company, in proportion to their investment. The government will contribute the spectrum resources to reduce the capital investment needed to deploy the network.

Under the scheme, the operators will buy capacity on the network at regulated wholesale prices. The open-access model means that in the future, other parties could buy capacity on equal terms to offer LTE services, even if they are not already participating. The initiative is also expected to contribute to the government's objectives for bringing broadband to schools, hospitals and government facilities.

Pros and Cons

Moving away from multiple competing mobile networks towards a shared wholesale network model where operators buy capacity from just one or maybe two wholesale providers would be a radical change from the structure of the mobile sector today. This may eventually happen everywhere, over a long period, or else manifest itself in only a limited way, where specific conditions dictate. For example, Mexico's 2012 decision to build a wholesale open-access network in 700 MHz is widely seen as an attempt to break the stranglehold of América Móvil, which has about 70 per cent of mobile subscribers.

If successfully implemented, a netco model could have a number of benefits: more efficient asset utilisation and lower costs for operators, faster and wider deployment of the latest technologies, greater spectral efficiency, lower tariffs and a wider choice of suppliers for end users. But in the worst case, it could result in an unwieldy, slow-moving and ill-managed monopoly network provider that stifles service competition.

Whether multiple or single networks are better for competition is a 'classic dilemma', according to regulatory economist Professor Martin Cave, author of the blueprint for spectrum liberalisation in the United Kingdom.* A merger, joint venture or input sharing may reduce costs in the short term, but restrict competition, encourage collusion and weaken competitive pressures to cut costs or innovate in the future, he said. In merger cases, the dilemma finds expression in the 'efficiency defence', Cave said. That is, 'let us merge because customers will benefit from cost savings'.

Mobile firms are fairly mature, market entry is difficult and operators produce similar outputs, Cave said. The same providers have gone through the mobile voice generations and are now switching to data services. 'This stability suggests to me a significant risk of tacit collusion', which could be facilitated by 'excessive' network sharing, he said. 'That's why I'm sceptical, and am inclined to draw the line with non-discriminatory passive sharing',† Cave added.

---

* 2002 Cave report for the UK Department of Trade and Industry.
† Passive sharing is the joint use of physical space, such as the site compound, masts and cabinets, air-conditioning and power supply. This is contrasted with active sharing, where parts of the RAN and electronics which make up the network are also shared.

Industry Views

The mobile industry association, the GSMA, is also sceptical about the benefits of wholesale networks. Although no single wholesale networks were in operation at the time of writing, the GSMA commissioned a report which compared roll-out rates from single networks against roll-out rates from multiple networks. 'The empirical evidence from more than 200 countries over a 15-year period shows that network competition has driven mobile network coverage for 1G, 2G and 3G networks further and faster than has been achieved by single networks. We would expect this to apply to 4G coverage too', said the report.

'[T]he traditional approach of licensing competing mobile operators can deliver more growth, innovation and societal benefits', said the GSMA's head of public policy, John Giusti, adding that they would be making this point to governments thinking of setting up single wholesale networks.

However, the UK's fixed line incumbent, BT, which bought national 2.6 GHz spectrum in 2014, is a supporter of wholesale networks. It argues that existing networks, even when upgraded with new technology and more spectrum, will be unable to deal with the capacity crunch. 'Building duplicate networks, with each assigned relatively small blocks of spectrum, is unlikely to be optimal. There are significant cost and performance benefits that can be achieved from shared infrastructure and spectrum. A neutral host network, assigned large blocks of spectrum, with regulated wholesale access to promote competition, could enable the required huge increases in network capacity to be delivered at lowest cost', BT said in a 2014 statement about policy in the UK market.[†]

Impact on Spectrum Liberalisation

As we have seen there are considerable pressures towards the sharing of mobile networks, and it is certainly possible – but by no means certain – that it may lead to the setting up of further wholesale LTE networks.

---

[*] Frontier Economics. (2014). *Assessing the Case for Single Wholesale Networks in Mobile Communications*, September, p. 3.

[†] Ofcom. (2014). *BT's Response to Ofcom's Consultation on Mobile Data Strategy*, January, p. 2 (http://stakeholders.ofcom.org.uk/binaries/consultations/mobile-data-strategy/responses/BT_Response.pdf).

In late 2014, these had been commissioned in Kenya and Mexico and were under consideration in Rwanda, Russia and South Africa.

If wholesale networks come to replace individual networks, then the current model for the liberalisation of mobile spectrum will need considerable revision. Mobile operators will no longer be competing to buy the most spectrum to provide the best services, they will instead be buying more capacity on a wholesale network. The netco running this network will be the one buying the spectrum. This could be managed through a liberalised approach, perhaps with competing netcos, but it seems likely that regulators would be concerned about the competition implications and insist that mobile operators had at least *some* of their own infrastructure.

The case for network sharing and even wholesale networks is clearly getting stronger, boosted by financial pressures, regulatory pressure to improve coverage and the technical characteristics of LTE. Being specific on how this will affect spectrum liberalisation is difficult because it depends on future developments and individual country circumstances.

However, we should note that these pressures could take the mobile sector into territory where the liberalisation model is either less appealing or requires substantial revision.

17

THE ROLE OF POLITICS IN SPECTRUM LIBERALISATION

Decisions about spectrum policy do not take place in isolation. They are made in the context of national political structures and institutional arrangements between states, these being principally exercised through the International Telecommunication Union (ITU). In this chapter we look at how these wider structures work and examine whether they are conducive to the introduction of spectrum liberalisation.

Considering the wide variety of national political structures, assessing whether these help or hinder the cause of spectrum liberalisation does not yield any simple answers. However, some themes do emerge. Not all political or legal structures are supportive of the principles of spectrum liberalisation: in this chapter, we focus particularly on problems in Thailand, Poland and Iran.

Another motif is that while a country may be generally in favour of developing spectrum markets, this can be a lower priority than other national objectives, such as balancing the budget or promoting the interests of a large national industry. We consider some cases where these latter objectives seem to have had a large influence on spectrum policy.

A further theme is that spectrum is highly prized as a national asset and governments are very reluctant to cede control of it. The European Commission has made credible arguments for the economic benefits of dealing with spectrum at a regional level, but attempts to pool sovereignty have been rejected by member states.

Nation States

Ultimately, spectrum is a sovereign matter for nation states. Not only do states have the monopoly over the legitimate use of physical force within a given territory, but they also have the right to govern the

177

use of the radio spectrum. Countries may sometimes delegate that authority to regional bodies, like the European Union, as we discuss later, but this does not detract from the nation state as the ultimate repository of power.

Strictly speaking, the use of all wireless equipment in a given territory is authorised by the state. When you use a mobile phone, it uses spectrum that your Mobile Network Operator has typically paid for in an auction. In effect, a part of your phone bill goes towards the state to pay for the cost of clearing that spectrum. When you watch over-the-air TV or listen to the radio, you use spectrum that was granted to broadcasters after the state deemed that they are suitable users of spectrum.

The state is also the sole body who has the right to enforce spectrum rights. To this end, it typically sets up an independent body that investigates complaints of interference from licence holders and enforces their spectrum rights. For example, in the Netherlands, the state agency Agentschap Telecom is considering fitting municipal garbage trucks with Mobile Data Collection units. The idea is that because each address is visited by a garbage truck every week or so, the regulator can build up a comprehensive database of radio-activity in a particular location, even if it is at low emissions. The regulator thinks this will allow it to detect unauthorised transmissions with great ease. The independent regulator also typically manages the assignment process for new spectrum releases, governs whether current licence conditions are being met, draws up long-term strategies for spectrum use and negotiates on behalf of the nation state at intergovernmental meetings.

Each state chooses to manage spectrum in its own way, and sometimes the portfolio of National Regulatory Authorities varies. According to UK law, Crown bodies, such as the Ministry of Defence, are not bound by Ofcom. In the United States, public sector spectrum is regulated by the National Telecommunications and Information Administration (NTIA), whereas private sector spectrum is regulated by the Federal Communications Commission (FCC).

The extent to which a regulatory body is independent from the government of the day also varies between different countries. In the United States, the FCC is governed by five commissioners, who are directly appointed by the US president and are subject to confirmation from the Senate. Writing in 2015, there were two Republican commissioners and three Democratic commissioners, including Chairman Tom Wheeler. The NTIA is an executive agency of the Department of Commerce and, again, is led by a presidential appointee who is confirmed by the Senate. By contrast, Ofcom acts according to the Communications Act 2003, and political decisions are left to the Department for Culture Media and Sport. Clearly, when institutions are subject to more political control, then they are also more subject to fluctuating political opinions.

Complicating the picture, governmental advisory committees sometimes lean into spectrum policy debates. These committees may work on an ad hoc basis, or may be permanent entities. The committees often have the power to shake up the direction of spectrum policy. In 2012, the US President's Council of Advisors on Science and Technology recommended that 1000 MHz of spectrum be shared between the public and private sectors. These proposals were seen as very controversial.

Radio waves are not only seen as a scarce national resource, but their use is often a matter of national security. Throughout the world, more spectrum is assigned to the military than any other type of user. In the United Kingdom, for example, 52 per cent of the entire radio spectrum is used by the public sector, and 75 per cent of public sector spectrum beneath 5 GHz is used by the Ministry of Defence. For national security reasons, military bodies generally avoid revealing how much spectrum they actually use, or in what way. Generally speaking, the armed forces are politically influential and governments tend to listen when the military speaks. Rival spectrum users struggle to engage with their arguments, partly because national security is always a government's first priority, and partly because there is little publicly available information to scrutinise.

The concerns of consumers, or more commonly, businesses wishing to safeguard access to spectrum, are articulated in a number of ways. Independent regulators frequently issue consultations on their latest proposals. Also, many companies come together to form interest groups. These groups vary enormously and can fit into many categories. They may be ad hoc groups that may disappear after an issue is resolved, they may be global organisations to whom spectrum is only one part of their work (GSMA), they may be representing a particular technology (Small Cell Forum) or a particular industry (Mobile Operators Association) or they may represent a broad range of industries with varying interests (UK Spectrum Policy Forum).

Individually or through industry bodies, companies may also take legal action to further their interests. In late 2014, a group representing the broadcasting industry in the United States, the National Association of Broadcasters, sued the FCC at the Court of Appeals for the District of Columbia over the terms of the planned 600 MHz band spectrum auction. Similarly, the UK's multiband 4G spectrum auction was delayed for many years due to numerous legal challenges from UK's wireless industry.

Sovereignty Can Be Complex

In some countries, the issue of sovereignty is not straightforward, and this has implications for spectrum policy. Whether Native American Indians have absolute or restricted sovereignty over their tribal lands in the United States has been a matter for legal debate, with some claiming that Native Americans have jurisdiction over the use of spectrum within those territories. This has allowed tribal lands to be used for trials of some spectrum technologies not allowed elsewhere in the United States.*

There is also a debate about sovereign rights in New Zealand. The country has a unique constitutional framework created when the early conflict between British settlers and the indigenous Maori people was settled by the 1840 Treaty of Waitangi. This treaty establishes the government's right to govern, its 'kawanatanga', but also established the Maori's sovereignty, or 'rangatiratanga', over 'taonga', or treasured things. Disputes over the interpretation of this treaty are settled by the Waitangi tribunal.

In 2000, the tribunal ruled that radio spectrum was 'taonga', and could not be auctioned by the New Zealand government. Strictly speaking, this should have prevented a planned 3G auction. However, as a compromise, the government agreed to reserve 25 per cent of the spectrum rights for a Maori trust at a subsidised price. The government also provided the Maori with some

---

* See Lessig. (2001). *The Future of Ideas*, Random House: New York, USA, p. 81.

cash that they could use to purchase the spectrum. Having acquired the spectrum rights, the trust entered into a commercial relationship with Econet Wireless and set up another competitor in New Zealand called 2degrees.

The issue resurfaced in 2013 when the government intended to auction the 700 MHz band for Long Term Evolution. The New Zealand Maori Council, the Wellington Maori Language Board and a third claimant reactivated a dormant proceeding about the matter at the Waitangi tribunal. Professor Whatarangi Winiata said, 'This is another confiscation. It's a 21st century confiscation and it's the nation that needs to say to the Crown, "we don't like what you're doing"'.

The tribunal, however, refused to grant an urgent hearing, and the government proceeded with its assignment. In reference to the right for Maoris to broadcast content in their own language, Telecoms Minister Amy Adams said, 'We have never accepted that spectrum is a taonga, but that the language is'. She added that 'We don't therefore accept that there's an automatic claim to spectrum'.

The auction concluded in Winter 2013, and the spectrum was assigned on an entirely commercial basis.

The Work of the International Telecommunication Union

The problem with national sovereignty over spectrum is that radio waves do not respect man-made borders. Deploying a high-power service such as broadcasting will prevent the same band being used for lower power services such as mobile in neighbouring countries. To make a fictitious example, if this were done by TV stations in Luxembourg, where the capital is less than 20 km from the borders with France, Germany and Belgium, it would knock out mobile services in substantial sections of those countries. Some international cooperation is required, and this is done through the ITU, specifically its Radio Bureau (the ITU-R).

The ITU is an agency of the United Nations and provides a framework for the international use of the airwaves by agreeing an intergovernmental treaty between states, known as the Radio Regulations. This is revised on an ongoing basis and the updated version agreed every 3–4 years at the World Radio Conference (WRC).

How Radio Regulations Work

The Radio Regulations work by categorising the uses of the airwaves into different types of services. The first subdivision is between space and terrestrial radiocommunications, but there are 40 different types of services in total. These include fixed services, the land mobile service, the broadcasting services, maritime mobile services, and the standard frequency and time signal service.* These types of services do not necessarily correspond with familiar uses of the airwaves. For example, mobile cellular phones are part of the land mobile service, but so is paging and the dispatch radio used by couriers.

However, the Radio Regulations do not stop countries deploying a particular type of service in a particular band: this is a common misconception. The regulations are an obligation *not to interfere* with services in other countries. To return to our fictitious Luxembourg broadcasters: if their frequencies were designated as a mobile band by the ITU, Luxembourg is not 'forbidden' to use it for TV. But, it would be breaking ITU treaty obligations if the broadcasting use *interfered* with mobile services in neighbouring countries. Luxembourg could use the band for broadcasting if it operated at such low power that the signal did not stray across borders, but this is unlikely to be cost effective as transmissions would be received in so few of the intended homes.

The Radio Regulations accept national sovereignty over spectrum: they are agreements *not to cause interference* in other countries. In practice, this means that most countries use bands in the manner indicated in the regulations.

Adhering to the Radio Regulations is made easier by splitting the world into three parts. ITU Region 1 is Europe, Africa and the Middle East; Region 2 incorporates all of the Americas; and Region 3 includes all of South and East Asia, as well as Oceania. The flexibility of the Radio Regulations is also increased by using the concept of primary and secondary allocations. This means that a band could be allocated to two services in a particular region, but the secondary users must deploy their services in such a way that there is no interference

---

* The World Bank's ICT Regulation Toolkit gives an excellent explanation of these categories. See the section called 'Radiocommunication Services ITU Regulation' (http://www.ictregulationtoolkit.org/).

Figure 17.1   ITU-R regions.

to primary users. It is also possible to have a co-primary allocation, where both services have equal priority in a band, and individual countries decide which service to use.

Footnotes are another aspect of the Radio Regulations and are used to make exceptions for particular countries, or to add a more detailed caveat (Figure 17.1).*

Strictly speaking, a spectrum band is *allocated* to a service by the ITU-R in the Radio Regulations, whereas nation states *assign* a band to a particular user. For example, in Region 2, the 698–806 MHz band has a co-primary allocation to mobile and broadcasting services, a secondary allocation for fixed services and a footnote next to mobile that indicates that it is a secondary allocation in Brazil. In the first quarter of 2014, the Chilean government decided to hold an auction to assign the 723–738 and 778–793 MHz bands to Entel, the 738–748 and 793–803 MHz bands to Claro, and the 713–723 and 768–778 MHz bands to Telefonica.

To summarise this section, the Radio Regulations are an extremely important determinant of whether a particular service, and therefore the technologies in that category, can have access to a particular band. The ITU working groups which study these issues and produce recommendations to be agreed upon at each WRC are therefore very important battlegrounds for technology manufacturers. The private sector can attend ITU meetings, but they do not have a vote at the WRC because these are intergovernmental meetings. The search for consensus drives work at all levels of the ITU, but there are some

---

* See Appendix A for more detail on the working of the ITU.

seldomly used voting methods available for national administration at WRCs if consensus cannot be found.

In a liberalised spectrum market, firms would buy spectrum and choose whichever technology would be most profitable. Ideally, companies would have a very wide choice, but in fact these choices are limited by the process of allocating spectrum at the ITU level. The ITU process tends to be influenced by manufacturers' priorities, but also by the promotion of national interests.

Spectrum as an International Bargaining Chip at the ITU-R Level

Let us consider an example of how international decisions on spectrum can be influenced by national political priorities.

The ITU-R Study Group 5 regularly updates ITU-R Recommendation M.1036, which governs frequency arrangements for International Mobile Telecommunications (IMT). Effectively the group decides whether the particular bands can be used for IMT, and which standardised technologies would be most suitable. The 1980–2010 MHz bands and the 2170–2200 MHz bands (part of the S-Band) have a global co-primary allocation for both mobile and mobile satellite in the band, but the band is not included in Recommendation M.1036 which deals with frequency arrangements for IMT. Effectively this prevents the band from being used for IMT. In 2014, South Korea proposed that the Recommendation should be expanded to include IMT in the S-Band. The proposal was opposed by China, citing the dangers of interference to MSS users. Because of the way the group votes on these decisions, China probably would not be able to ultimately prevent the adoption of a band-plan for the IMT use of the band, and would in any case still be able to assign the band domestically according to its own needs.

However, in Autumn 2014, the French regulator decided that it would support China's position, and oppose the inclusion of the S-Band in Recommendation M.1036's upcoming amendment. At first glance, this seems like a counterintuitive move for an administration which is generally supportive of the telecoms industry. However, a political reading of the decision suggests that this move is connected to France's public deficit.

Allegedly, the French administration government desperately needs to fill the shortfall in its military budget, so it has pushed forward with plans to assign the 700 MHz band, which it hopes will raise at least €2 billion. It now plans to assign the band by the end of 2015, despite the administration's assurances in 2011 during the 800 MHz band auction that no more UHF frequencies would become available before 2020. However, one barrier to the 700 MHz band auction is that the CEPT's preferred band-plan for the spectrum has not yet been added to M.1036, and the French administration is worried that the inclusion of the S-Band will risk the amendment not being passed. Conversely, the Russian administration supports the South Korean position despite its traditional opposition to the use of spectrum for IMT. It has been privately alleged that Russia's position gives it a bargaining chip that can be used amid wider political negotiations about spectrum policy.

Spectrum and National Politics

Spectrum Auctions as a Tool for Plugging Budget Deficits

Spectrum liberalisers argue spectrum should be sold at the price the market is willing to pay. If the auction is manipulated to force prices up, then the winning companies may go bust and the nation will be the greatest loser, missing out on job creation, increased economic efficiency and improved tax revenues. However, cynics argue that the amount of money generated in spectrum auctions makes them an irresistible 'cash cow' for revenue-starved governments, incentivising the manipulation of rules to maximise income.

Following the telecoms crash in the early 2000s, the amount of revenue that governments raise in auctions has declined. The UK 3G spectrum auction in 2000 raised £22.47 billion for the UK Treasury. By comparison, the UK's 800 MHz and 2.6 GHz bands '4G' auction in 2013 raised only £2.34 billion. Taking into account inflation (assuming the 2000 auction raised the equivalent of £34 billion), the 2013 auction raised only 6.9 per cent as much as the 2000 auction.

A few months later in 2000, the German regulator held a large multiband auction of spectrum in the 1.9 and 2.1 GHz band. The auction raised the equivalent of €50.5 billion (DM 98.8 billion). In 2010, an auction of spectrum in the 800 MHz, 1800 MHz and 2.6 GHz bands only raised €4.4 billion, substantially less than the amounts single lots raised in 2000.

By 2013, the amount of revenue that governments raised in spectrum auctions had substantially fallen. In its budget predictions the UK government had relied on the 4G auction to raise £3.5 billion. The opposition's telecoms spokesperson criticised Ofcom for not raising more money. 'It's quite wrong that in a time of austerity the mobile phone companies were given spectrum at prices below even what they were prepared to pay', she said.

This short-lived wave of high-revenue auction has been mocked by the highly respected academics and consultants, Gerard Pogorel and Erik Mohlin. In a recent paper, they write, 'Spectrum auctions were designed at a time of financial exuberance, loss of regard for the realities of the productive economy, blind faith in narrowly defined market mechanisms [and the] use of fiscal expedients to cope with always deeper imbalances in public spending'.

Following the telecoms crash in the early 2000s, the amount of revenue that governments raise in auctions has declined. Nevertheless, the Australian 700 MHz auction in 2013 was widely seen as an attempt to maximise revenue. After the telecoms minister, Stephen Conroy, was privately told that Vodafone would sit out of the auction, he decided to overrule the independent regulator, ACMA, and set the reserve price himself. A 2 × 15 MHz block was left unsold and the auction raised $1 billion AUD ($811 million USD) less than the government had predicted.

How National Political Structures Influence Spectrum Policy

This book talks about the attempts to apply the principles of market liberalisation to spectrum, and clearly that process will be easier if a country's economy in general is more liberalised.

Here we should note briefly that free market principles are associated with a particular political approach, often labelled *Anglo-American Capitalism*. This is dominant in the United States and the United Kingdom, but is also found in other English-speaking countries like Canada, New Zealand, Australia and Ireland. It emphasises free enterprise and private ownership over government planning and nationalised industry. It is often contrasted with the more corporatist approach to capitalism found in Europe and Scandinavia. This is satisfied with state intervention and tends to have enhanced collective bargaining rights for workers and a greater commitment to the welfare state.

The accuracy of the 'European Corporatism versus Anglo-American Capitalism' distinction is much debated, but for our purposes it is a useful entry point. Since the 1980s, most countries in Europe have moved more towards the Anglo-American model by selling off nationalised industries. The privatisation of national telecoms companies, which often have mobile arms, has been a key part of this process.

Anywhere in the world, where there remains state ownership of industries that use spectrum, the implementation of spectrum liberalisation is less straightforward.

Thailand

Thailand is an example of a country when mobile spectrum is held by the state, but this is done through an unusual structure, known as the *concessionary system*. Mobile spectrum belongs to two state-owned telcos, TOT and CAT Telecom, who then lease a spectrum 'concession' to AIS, TrueMove and Digital Phone Co (DPC) with the licensing system that is common in the West.

The country's independent spectrum regulator, the National Broadcasting and Telecommunications Commission (NBTC), was trying to change the country's spectrum policy from to a licence-based system but this was interrupted when the Thai military performed a coup d'etat in May 2014, and set up the National Council for Peace and Order (NCPO) to govern the country.

Part of the NBTC's reform process was two auctions that were to be held in Summer 2014: an auction of 17.5 MHz in the 900 MHz band and another auction of 25 MHz in the 1800 MHz band. However, the military retains close links with the state-owned telcos and AIS has links with ousted Prime Minister Yingluck Shinawatra. Her brother, Thaksin Shinawatra, founded the company in 1986, before going into politics and becoming prime minister in 2001. He tried to privatise the two state-owned firms but was himself deposed in a military coup in 2006. He now lives in Dubai.

Upon assuming power in May 2014, the NCPO immediately suspended NBTC's plans to hold both auctions, pending an 'investigation'. In August, the NCPO told the NBTC that it had to redesign the rules for the contest and would have to hold the auction a year later. The NBTC told the press that it was considering holding a beauty contest for the spectrum instead of an auction. Meanwhile, TOT asked the NCPO if it can keep its spectrum for another 15 years.

Thailand is an example of how efforts towards spectrum liberalisation can be clouded by the associations between the state, successive governments and the mobile industry.

Poland

The tools which would routinely be used to apply market mechanisms to spectrum are not always present in national legal frameworks. A good example of this is the '4G' auction in Poland. During planning for this in 2014, it emerged that the rules normally used in auctions were incompatible with Polish law. Participants in Polish auctions were able to make bids and then withdraw them, even if they won the auction outright. This is usually forbidden in spectrum auctions because it would allow companies to take part in the auction solely for the purpose of pushing up costs for competitors. The Polish regulator also lacked the legal instruments to check that bidders were not merely 'front' companies for an existing participant, which can also be used to manipulate the auction result.[*]

During the debate on the auction rules, it was argued that all the necessary legal instruments were present in EU law and could therefore be used in Poland. That was not the regulator's interpretation, but whether that is correct or not the predominant business and political culture in Poland was clearly holding back the application of the principles of spectrum liberalisation.

Iran

The jamming of satellite signals in Iran is a further illustration of spectrum liberalisation depending on a wider framework of laws which guarantee human rights and property rights.

---

[*] See Version 3 of Polish 4G Auction Framework Baffles Operators, *PolicyTracker*, (2014) (https://www.policytracker.com/headlines/version-3-of-polish-4g-auction-rules-leaves-operators-nonplussed).

The Islamic Republic of Iran has been widely accused of jamming satellite signals since 2003 in order to prevent domestic exposure to Western media. This is undermining the property rights of Intelsat, and Eutelsat, who use spectrum allocated to them globally in the Master International Frequency Register. In many counties, it would also breach free speech legislation.

The Iranian government denies any involvement in jamming. However, Iranian dissident diaspora groups find this claim absurd. They point out that satellite operators have pinpointed the source of interference to territory in Iran and that the approximately 120 Persian-language satellite TV channels that are transmitted from outside of Iran are all domestically banned. This ban is enforced through raids on private property. Further, anecdotal evidence suggests that jamming occurs more frequently during BBC Farsi-language news broadcasts and in times of political uncertainty, particularly during the contested presidential election in June 2009.

After strong lobbying by the French and British governments, the WRC-12 agreed to amend a part of the Radio Regulations, making governments responsible for any jamming activity in their jurisdiction. In other words, the Iranian government has an obligation to track down the jammers and prevent their work.

The amendment to the regulations had little effect until Iran elected a reformist president, Hassan Rouhani, in September 2013. The new president has a less hardline approach and even said that 'in the age of digital revolution, one cannot live or govern in a quarantine'. Rouhani's actions are limited by the country's constitution, which places many checks and balances on the president's power among the religious establishment. In February 2014, Rouhani ordered the Health Ministry to investigate whether or not jamming is linked to reports of cancer. The pressure to change the unofficial policy grew in July when signal jammers were blamed for the failure of the domestic meteorological office to predict a major dust storm that hit Teheran a month before. Five people died in the storm.

Whatever the outcome of the public health investigation, it is clear that changes to wider political cultures have an influence on efforts to liberalise spectrum.

Spectrum and Regional Politics

The EU Radio Spectrum Policy Programme

The European Union is widely regarded as the most successful international single market (i.e. an area where people, goods, services and capital can move from one country to another without borders, to stimulate competition, prosperity, and trade and improve efficiency). This achievement has its roots in the 1957 Treaty of Rome and found more concrete implementation in the Single Market Act of 1986, but these milestones hide deeper disagreements. There has always been conflict about how a single market should be defined and how it should be achieved in practice.

Signatories to the Treaty of Rome (all EU member states) make a high-level commitment to seek 'ever-closer union', but what this means in terms of practical measures remains a divisive issue. In the broadest sense, some pro-integrationist European politicians are happy for EU institutions to have increased power. Others want nation states to retain as much power as possible with the EU functioning more as a free trade area.

Pooling sovereignty is unpopular in many EU countries: in the 2014 European elections, Eurosceptic parties topped the polls in both the United Kingdom and France. In the same year, about a third of Europeans (32 per cent) said their country would be better off outside the EU, according to Brussels' own regular survey.[*]

Spectrum policy has become a pawn in these pro- and anti-integrationist arguments, and this started with Global System for Mobile (GSM), arguably the world's most commercially successful wireless communications technology. It first took off in Europe, and its success has widely been attributed to the European Commission's decision to mandate all member states of what was then the European Economic Community to reserve the 900 MHz band for GSM. This was done through the 1987 GSM Directive, which obliged member states to not only reserve the spectrum, but also to issue licences for the use of GSM in that territory.

But for some would-be supra-nation builders, the obvious question was why stop there? Why not harmonise all spectrum? This was

---

[*] Eurobarometer 415, *EUROPEANS IN 2014*. European Commission, July 2014.

the line of thought behind the Communications Framework Review, which was published in June 2006. Presenting the paper, Information Society Commissioner Viviane Reding emphasised the importance of creating a 'single market for spectrum'. She said, 'It is a competitive disadvantage for Europe that we do not have, as in the US, a single regime for spectrum management, but 25 different ones. That is why many stakeholders propose the creation of a European spectrum agency. I believe we should discuss this idea'.

In the European Union, both the Council of the European Union and the European Parliament have to approve an initiative from the European Commission, which is considered to be like an executive. The Council is made up of the relevant ministers from each of the EU's member states, whereas the Parliament is elected in regular pan-European elections. The Council is considered intergovernmental but the Parliament is considered transnational. The Commission's proposals to centralise spectrum policy making in Europe have been rejected by both the Council and the Parliament.

By November 2007, Reding's ideas had evolved into a draft proposal to set up a new regulatory body known as the European Electronic Communications Market Authority (EECMA). At the time, it was reported that the majority of national regulators, legislators and industry officials opposed the spectrum aspects of the plan. By 2008, concerns about the growing power of the European Commission led the European Parliament to block plans to liberalise the GSM bands, even though it was theoretically in favour of doing so. The Council also opposed the EECMA.

Over 2008, a new consensus emerged that EECMA should be replaced by a much diluted advisory organisation known as the Body of European Regulators in Telecom, which later became the Body of European Regulators for Electronic Communication. They became a part of the Telecoms Package, which itself got dragged into a discussion about how to deal with Internet piracy. The French government wanted to introduce a 'Three Strikes' process that would have blocked pirates from accessing the Internet after a third copyright breach, but this was opposed by the majority of MEPs. Regarding spectrum, the new framework allowed for a compulsory 'road map' for spectrum for all member states to follow, and no change in spectrum competence between the European Commission and member states.

In February 2012, some 6 years after the Commission first flew a kite on a single spectrum regulator, the European Parliament approved the Radio Spectrum Policy Programme (RSPP). The RSPP does contain some key conditions of spectrum liberalisation. For example, spectrum trading must be enabled by 2015 in all of the European Union's 'harmonised bands'. These harmonised bands are the GSM bands that have been harmonised since 1987, along with the 3.4–3.8 GHz band, the 800 MHz band and the 2.6 GHz band. Each of these bands also had to be assigned by all member states within certain deadlines. As mentioned in Chapter 13, the Commission was also tasked to create an inventory of spectrum usage that would guide future policy making.

In September 2013, Vice President of the European Commissioner Neelie Kroes who was responsible for the Digital Agenda presented a package of legislation designed to enable the completion of the digital single market. The announcement came shortly after she publicly expressed her exasperation at the majority of member states' failure to assign the 800 MHz by 1 January 2013, as was agreed in the RSPP.

Documents leaked prior to the launch showed that the Commission was considering re-introducing single European licences, but these provisions were absent from the final proposals. European stakeholders generally had mixed views on this new package. An amendment by the European Parliament created a provision that some thought would further the cause of spectrum liberalisation in Europe – such as minimum licence durations of 25 years. However, Kroes did not manage to get the proposals adopted by the time her term expired in November 2014, and documents surfaced in January 2015 suggest that the European Council will remove all spectrum measures from the legislation.

It is hard to say whether attempts to have a unified approach to spectrum policy across the EU have helped or hindered spectrum liberalisation. In one sense, a pan-European market for spectrum could be economically beneficial, reducing costs for operators. In another sense, it could restrict the ability of national markets to respond to their own individual circumstances. In general, the European Commission has been in favour of liberalisation, backing attempts to introduce trading and generic licences. But it also tried to mandate the use of the DVB-H mobile TV standard – an attempt to bypass market mechanisms and favour a European technology.

What is clear from this discussion is that European Commission initiatives on spectrum policy have been closely linked with a pro-integrationist agenda where more power is passed from nation states to EU institutions. Have Brussels' spectrum proposals been judged by the company they keep rather than on their own merits? Perhaps so, and this contributes to the detriment of the spectrum liberalisation cause.

Conclusions

This chapter has considered how wider political structures impose further constraints on the invisible hand that spectrum liberalisers hoped would allocate scarce spectrum as efficiently as possible.

We have seen examples of how countries' national financial imperatives have influenced international negotiations about spectrum policy and also national policies. In both the European Union and Iran, we have seen how spectrum policy has become inextricably linked to a wider political agenda. Overall, it is certainly possible for wider national and international political constraints to exert a negative influence on spectrum policy reform.

18
CRITICISMS OF AUCTIONS

As discussed in Chapter 4, auctions have become the most popular part of the liberalisation agenda, adopted almost worldwide, even in countries that are nervous about other market mechanisms.

The delays and legal challenges associated with beauty contests, compared with the transparency of auctions as well as the economic arguments have seen the latter become the default assignment mechanism. However, in recent years cracks have appeared in this pro-auction consensus, with the European Commission the highest profile doubter. Some are predicting a move back to administrative assignment.

Many of the criticisms of auctions which gained new currency in 2010 onwards had been made during the debates which precipitated the move from beauty contests to auctions in the 1990s. They have been raised again for two reasons: the first being concern about investment in new mobile networks. This is a particular worry in Europe, where operators have claimed huge spectrum fees prevent them updating their networks.

The second reason is frustration with the oligopoly which has often come to dominate the mobile sector. What started life as a dynamic sector with many more competitors than its fixed line counterpart has ended up as a staid market with only three or four players. In the eyes of some observers, auctions are contributing to this unhealthy state of affairs.

In this chapter we concentrate on three criticisms of auctions:

1. Do they discourage investment?
2. Do they hinder competition?
3. Is the increasing complexity of auctions potentially damaging?

We then set these criticisms within the wider debate about liberalisation and consider how influential they are likely to be.

Do Spectrum Auctions Discourage Investment?

Concerns of the European Commission

By the second decade of the century, Brussels was becoming concerned about the region losing its lead in the telecoms market. Formerly dominant manufacturers like Nokia and Alcatel-Lucent were in decline and Long Term Evolution (LTE) roll-out was progressing much quicker in the United States. This led to concern about the generally higher prices paid in EU spectrum auctions leaving the industry with no money to invest in new networks. Investment in 4G was crucial if Europe was to make up lost ground and recover from the 2008 financial crash which sent many member states into recession,[*] argued former Digital Agenda Commissioner Neelie Kroes in 2013.

She was particularly critical of the 2012 Dutch mobile licences auction, which raised €3.8 billion, eight times more than expected and four times more per capita than in Germany. 'Was nothing learned from previous auctions for UMTS frequencies, when the share price of KPN [Dutch incumbent] dropped substantially and the ecosystem of small supply companies in the telecom sector was severely damaged?' she asked. 'The government absorbed all the money before even a single Euro was earned by the new mobile phone and internet services'.

The Dutch auction also got a negative reaction from investors: shares in KPN plummeted by 14 per cent when the markets opened the business day after the auction ended. Two other winning bidders, Vodafone and multinational German group T-Mobile, also saw their share prices drop.[†]

Although Kroes' concern is understandable, it is not clear exactly who or what she is criticising. Is she blaming the Dutch operators for paying too much? If so how could that be stopped?

Or is she blaming the regulator for poor auction design? The auction followed the Combinatorial Clock Auction (CCA) design, which actually uses a second-price rule, as explained below. If a different

---

[*] Published on 11.1.13 in *Het Financieele Dagblad* in the Netherlands but available in English here: http://ec.europa.eu/commission_2010-2014/kroes/en/content/christmas-present

[†] *PolicyTracker*. (2012). Dutch auction results reveal contrasting LTE strategies. https://www.policytracker.com/headlines/dutch-auction-results-reveal-contrasting-lte-strategies

design using a first-price rule had been used, the amount paid by the operators is likely to have been higher.

Furthermore, the Dutch CCA contained a reservation for new spectrum, a feature encouraged in the EU's own Radio Spectrum Policy Programme (RSPP).* The high prices seemed to be generated by the arrival of a new entrant – Tele2 – and by the fact that nearly all mobile licences were being sold at the same time. If operators had not taken part, they would have been out of the game – there was no sitting this one out, as they could have done if only a single band were for sale.

Or was she saying that the Dutch government should not have taken all the money? If so, how would this work? Again there is no explanation.

Kroes' comments also suggest that there should have been an administrative assignment of the licences. 'Was nothing learnt from previous auctions for UMTS frequencies?' she asks, alluding to the 3G auctions in 2000, where bidders paid far too much. However, this issue has been much debated since, and the balance of evidence points to particular circumstances, rather than the use of auctions caused the overpayment.

Do Auctions Cause Overpayment?

No one would dispute that the spectrum at 2 GHz sold in Europe (commonly referred to as the '3G licences') in 2000 were overvalued. BT said its UK 3G licences had lost 75 per cent of their value between 2000 and 2005. In 2003, O2 wrote down the value of its UK licence to 47 per cent of the initial cost, and a Danish 3G licence was returned and re-auctioned in 2005 for 56 per cent of original value.[†] The question is, did auctioning cause the overvaluation?

The first weakness in this argument is highlighted in Figure 18.1. The biggest valuations were in the United Kingdom, the Netherlands, Germany and Italy and occurred between April and October 2000. True, these were all auctions, but there were also 3G auctions held in

---

[*] See Radio Spectrum Policy Programme (RSPP) published by the European Commission and available here 2 b) http://eur-lex.europa.eu/legal-content/EN/TXT/PDF/?uri=CELEX:32012D0243&from=EN

[†] French, R. D. (2009). *Governance and Game Theory: When Do Franchise Auctions Induce Firms to Overbid? Telecommunications Policy*, Vol. 33, Issues 3–4, p. 4.

Country	Date of assignment	Method	Licence fee/ population (UK = 100)
Finland	March 1999	Beauty contest	0
Spain	March 2000	Beauty contest	2
UK	April 2000	Auction	100
The Netherlands	July 2000	Auction	29
Germany	August 2000	Auction	97
Italy	October 2000	Auction	33
Austria	November 2000	Auction	16
Norway	November 2000	Beauty contest	1
Portugal	December 2000	Beauty contest	6
Sweden	December 2000	Beauty contest	0
Switzerland	December 2000	Auction	2
Belgium	March 2001	Auction	7
Greece	July 2001	Auction	7
France	July + December 2001	Beauty contest	3
Denmark	October 2001	Auction	15
Luxembourg	March 2002	Beauty contest	0
Ireland	August 2002	Beauty contest	15

Figure 18.1   European 3G spectrum assignments in chronological order. (Adapted from *3G Mobile Telecommunications Licenses in Europe: A Critical Review.* Harald Gruber, *info*, Vol. 9, No. 6.)

Switzerland, Belgium and Greece from December 2000 to July 2001, and these produced low values.

This points the finger of suspicion not at auctions themselves, but at a period of time, namely, the middle of 2000. This was the height of the dot-com boom, with the technology-heavy Nasdaq 100 stock market index peaking in April 2000, remaining reasonably high until the autumn, before going into freefall (Figure 18.2). And it was in autumn 2000 that spectrum auctions prices also started to fall.

These were extraordinary circumstances. The stock market (wrongly) believed that this was the only opportunity to buy spectrum for the mobile Internet. Any mobile company that did not have this spectrum did not have a future and its stocks would be punished accordingly. 'If the [operators] don't secure a license regardless of their price, the stock market decimates the company; if they win, the company bleeds itself over the license's lifetime (usually 15–20 years) as it struggles to make

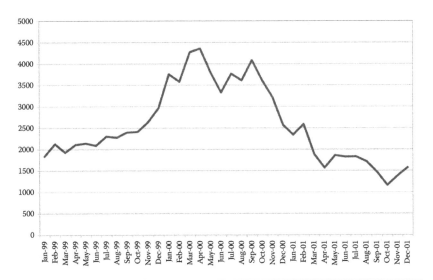

Figure 18.2   The Nasdaq 100 index 1999 to 2001. Note that the graph shows the opening price of the Nasdaq 100 index on the first day of each month. (Data provided by Investing.com.)

a profit', said Martin Bouygues, CEO of Bouygues Telecom in France. For operators, it was damned if you do and damned if you don't.

Uncertainly was another big factor in the valuations. The mobile Internet was new, nobody knew how much people would pay for it, and there was no market data to steer by. People involved in the bidding process said the back-room teams were asked to revise their business plans as the prices went up – this explains why.

French (2009)* helpfully points out that there is a precedent for uncertainty producing overvaluation. In India in 1995, operators offered a massive $6.25 billion for 34 licenses in a sealed bid auction, leaving them unable to meet the royalty payments to the government. Luckily, the government was able to cancel the payments to stop the companies going bust, and French points out that this is a little-appreciated advantage of not demanding up-front payments. The valuations had been based on an overestimate of an unknown: the likely spending power of India's emerging middle class. Like the mobile Internet, there was no reliable comparative data.†

---

* French, R. D. (2009). *Governance and Game Theory: When Do Franchise Auctions Induce Firms to Overbid? Telecommunications Policy*, Vol. 33, Issues 3–4.
† Ibid., p. 5.

Does the Regulator Know Best?

Returning to Kroes, she may have been trying to encourage support for greater Commission oversight of spectrum auctions. In 2011 and 2012, the Commission had been promoting a plan which would have led to synchronised spectrum auctions in several countries, with the aim of reducing the cost of multicountry licences and encouraging pan-European mobile services. The proposal would also have made national auction proposals subject to peer review by the member states.[*] In the end, these ideas did not form part of the Commission's RSPP.

Underlying Kroes' concerns about auctions is an argument which sounds right: the more you charge mobile operators for licences, the less cash they will have to upgrade their networks. If prices had been set by regulators, rather than being subject to the vagaries of an auction, the more moderate levels would have given more spare cash for investment.

However, the reality seems more complex. A study in December 2005 for the Swedish regulator, PTS, examined the progress in 3G roll-out achieved in most of the countries which had by then awarded licences.[†] By this date, Sweden had achieved the best roll-out in Europe, reaching 90 per cent of the population. The operators were paying no up-front licence fees, just 0.15 per cent of turnover (Figure 18.3).[‡]

This tends to support the idea that lower fees encourage investment. However, the second-best roll-out in Europe was the United Kingdom, where operators had paid an eye-watering total of £22.5 billion (then $35 billion) or 2.5 per cent of GDP for their 3G licences.

In global terms, the best roll-out (100 per cent) had been achieved in Japan, which had a beauty contest but the second best with 99 per cent

---

[*] *PolicyTracker.* (2013). Leaked document suggests Brussels spectrum power grab. https://www.policytracker.com/headlines/european-commission-plans-to-take-on-extra-spectrum-assignment-powers?searchterm=RSPP+auction

[†] Stelacon consultancy. (2005, December). *UMTS Development – From an International Perspective* (Published by Swedish Post and Telecom Authority (PTS) in Sweden and available here: http://www.pts.se/upload/Documents/EN/UMTS_development_Appendix_2_060309_06-04.pdf).

[‡] Björkdahl, J., and Bohlin, E. (2001). *Financial Analysis of the Swedish 3G Licensees*: Where are the profits?, *info*, Vol. 4(4), pp. 10–16.

Country	Date of assignment	Assignment method	Population coverage by 2005 (%)	Date coverage achieved
Australia	Mar-01	Auction	54	August 2005
Austria	Nov-00	Auction	62	June 2005
Denmark	Oct-01	Auction	60	December 2004
Finland	Mar-99	Beauty contest	30	October 2005
France	Jul-01	Beauty contest	60	October 2005
Germany	Aug-00	Auction	60	June 2005
Great Britain	Apr-00	Auction	85	March 2005
Greece	Jul-01	Auction	46	December 2004
Holland	Jul-00	Auction	60	September 2005
Hong Kong	Sep-01	Auction	99	August 2005
Ireland	Aug-02	Beauty contest	62	Date not given
Israel	Dec-01	Auction	92	Date not given
Italy	Oct-00	Auction	78	April 2004
Japan	Jun-00	Beauty contest	100	October 2005
Norway	Nov-00	Beauty contest	76	June 2005
Portugal	Dec-00	Beauty contest	60	Date not given
Spain	Mar-00	Beauty contest	70	Date not given
Sweden	Dec-00	Beauty contest	90	February 2005
Switzerland	Dec-00	Fixed price	90	April 2005

Average population coverage by 2005 (%)	
For all assignment types	70
For auctions	70
For beauty contests and fixed price assignments	71

Figure 18.3 Population coverage achieved after 3G auctions. (From Stelacon. (2005). *UMTS Development – From an International Perspective* with additional material by 3G Mobile Telecommunications Licenses in Europe: A Critical Review. *Harald Gruber, info.* Vol. 9, No. 6. With permission.)

was Hong Kong, which had an auction. The average roll-out achieved for any assignment type was 70 per cent, for auctions it was the same, and for beauty contests and fixed price assignments it increased by just 1 per cent to 71 per cent.

These figures suggest little correlation between high auction fees and low investment. One explanation may be that competitive markets produce high prices in auctions, but these same competitive markets mean that if one operator launches a new service – like 3G or 4G – the others will lose customers if they fail to make a similar investment.

Competition in mobile markets may not be optimal, but it is reasonable enough to force most players to upgrade their networks. It is only natural for mobile operators to lobby for lower licence fees, but any claim of a social benefit from reducing these charges needs to be examined closely.

Do Auctions Hinder Competition?

In theory, the business which can get the most value from the spectrum will be the one prepared to pay the most. The highest bidder will have the best business model, better marketing, better technologies and better management than their competitors. But is this the only explanation for auction prices? Recently, there have been suggestions that this purest market mechanism can obstruct competition, rather than open the market to new entrants.

Consultants who value spectrum prior to auctions say privately that companies put a value on preventing their competitors from accessing a particular frequency. Companies will first calculate the value of the spectrum to their business, then consider the additional value of preventing their competitors from getting it, or even the greater value of denying a licence to a new entrant. 'In many cases', wrote Stefan Zehle of Coleago Consulting in 2013,* this was 'the largest source of value' (i.e. a large part of the value of spectrum to a bidder has nothing to do with technical value or the ability to deliver better services but with reducing competition).

The mobile oligopolies bring out this behaviour in operators, even leading them to refuse to sell unused spectrum for fear of benefitting a competitor, according to another (anonymous) consultant.† The US Department of Justice anti-trust division also made this point in a 2013 submission to the Federal Communications Commission (FCC). "[C]ompetitors' lack of spectrum may require higher capital expenditures, such as having to build more cell towers, in order to provide a competitive service. Thus, a large incumbent may benefit

---

* *PolicyTracker.* (2013). Opinion: The end of spectrum auctions? https://www.policytracker.com/headlines/the-end-of-spectrum-auctions

† *PolicyTracker.* (2012). Market forces are not always the answer. https://www.policytracker.com/headlines/market-forces-are-not-always-the-answer

from acquiring spectrum even if its uses of the spectrum are not the most efficient, if that acquisition helps preserve high prices".*

The US Department of Justice argues that auctions may be the best assignment method in new markets but that becomes more questionable as the market matures. The solution they propose is to make more spectrum available which would allow new entrants to challenge the existing incumbents.

The question here is whether the importance of 'competition blocking' is being factored into the policy debate. Spectrum may go not to the bidder that values it the most, but to the bidder that places the greatest value on keeping out competitors.

In the worst-case scenario, this argument suggests governments are extracting large sums from the telecoms industry at auctions, partly because mobile operators put a high value on the spectrum, and partly because they also put a higher value on maintaining an oligopoly.

Goodbye to Auctions?

Zehle argues that auctions are now subject to a plethora of rules including 'band specific or overall caps, band specific obligations, limitations to bid based on market share, high reserve prices, roaming rules, deployment rules'† that negate the purpose of an auction, 'namely to allocate spectrum to the highest bidder'.

This goal is also obstructed by the governments setting high reserve prices which make auctions close within a small number of rounds, as happened in Greece in 2011.‡

'Sooner or later', Zehle says, 'regulators will abandon the dogma of auctions and accept that the industry is heading for consolidation,

---

* See *Ex Parte Submission of the United States Department of Justice 11.4.13. In the Matter of Policies Regarding Mobile Spectrum Holdings*, p. 11 (http://apps.fcc.gov/ecfs/document/view?id=7022269624).
† One of the architects of spectrum liberalisation in the United Kingdom, Professor Webb, is also critical of coverage obligations, though not of auctions specifically. See *PolicyTracker*. (2012). Market forces are not always the answer. https://www.policytracker.com/headlines/market-forces-are-not-always-the-answer
‡ *PolicyTracker*. (2011). Greek operators pay high price for 900 and 1800 MHz spectrum. https://www.policytracker.com/headlines/greek-operators-pay-high-prices-for-900-and-1800-mhz-spectrum

at least at network level, and may devise administered spectrum allocation mechanisms that "distribute" new spectrum among a reasonable number of operators, perhaps three or four in each market, depending on absolute size".* However, Zehle is concerned that governments tend to overestimate the optimum number of competitors, leading to a 'destruction of value'.

To summarise, Zehle predicts that the flaws in actions will lead to their demise but is uncomfortable with their likely replacement.

Licensed Shared Access to the Rescue?

Gerard Pogorel and Eric Bohlin, professors from France and Sweden, respectively, use the criticisms of auctions[†] to argue for the possibilities of Licensed Shared Access (LSA). (See Chapter 20.) They agree that the auction system has created semicompetitive mobile oligopolies: 'the repeated assignment of frequencies over a decade to the same bunch of 3 or 4 companies in each domestic market is not what we would call the image of a competitive market process'.[‡]

Governments have become so dependent on auction revenues that 'this has created the new version of the capture of the regulator'.[§]

Pogorel and Bohlin argue that LSA can be used to encourage new entrants, stimulate competition and create innovative goods and services if it is assigned via bidding for rents or royalties, rather than up-front fees as in nearly all current auctions.

The Problems with Complex Auctions

Auction formats have become increasingly complex in recent years in an attempt to overcome the limitations of previous formats, and the best known example of this complex format is the CCA. (Figure 18.4 summarises the pros and cons of the main auction types.)

---

* See note 39 for source.
[†] *Valuation and Pricing of Licensed Shared Access* (https://webperso.telecom-paristech. fr/front/frontoffice.php?SP_ID=9&).
[‡] Ibid., p. 16.
[§] Ibid., p. 16.

Auction type	Characteristics	Advantages	Disadvantages
First price sealed bid	Bids are submitted confidentially Highest bidder wins and pays bid amount	• Encourages new entrants	• No price discovery • Winners curse
Second price sealed bid	As above but winner pays second highest bid	• Avoids winners curse • Encourages people to bid their value	• No price discovery • Potential public relations disaster!
Proxy auctions[i]	Someone submits a bid on your behalf e.g. with eBay a computer bids to a specified amount	• Simple • Widely used	• Bidding in the last few seconds of a round (Sniping)
Simultaneous multiple round auction (SMRA)	Participants can bid on several items in each round, the price increasing with each round. Stops when no further bids are received	• Price discovery • Simple format	• Aggregation risk
Combinatorial clock auction (CCA)	Bidders can submit multiple bids on different combinations or 'packages' of licenses	• Price discovery • Avoids aggregation risk	• Potentially very complex • Will the public (and bidders!) understand? • Price determination can involve very advanced maths!

[i] This type of auction has only been used occasionally for low value spectrum, e.g. in Norway and Denmark in 2010.

Figure 18.4 The advantages and disadvantages of different auction types.

CCAs are replacing Simultaneous Multiple Round Auctions as the design of choice for auctions in Western Europe, but concerns remain about their complexity, transparency and ability to produce unexpected results.

First- and Second-Price Auctions

One of the defining features of the CCA is that it is a second-price rather than a first-price auction. This means that the winner is the person who offers the most, but they pay the price of the next lowest bid. To take a simple example, if we had the bids for a single item shown in Figure 18.5, A would win but pay the second price (i.e. 60), the amount bid by *B*.

Bidder	Price
A	100
B	60
C	55
D	45
E	30
F	20

Figure 18.5 Simple second-price auction.

The problem with first-price auctions is that the highest bidder can suffer *the winner's curse* (i.e. not be able to make a profit because of overpayment). In fact, the winner's curse can be seen as a structural problem inherent in first-price auctions.

You only win the auction if your value is higher than everybody else's, but if that is the case you may have overbid! Auction theorists have argued that this prevents participants bidding their true value and to resolve this problem looked back to academic work conducted in the early 1960s by the economist William Vickrey, who later won the Nobel Prize.

Vickrey argued that adopting a second-price principle encouraged people to bid their true value because they would pay not the amount they bid, but the value below that. This principle has been adopted into the design of the CCA.

Package Bidding

In the CCA, participants are bidding on packages of items, to remove the flaw inherent in earlier auction types, whereby bidders could win some but not all of their desired items (known as *aggregation risk* or *exposure risk*). The CCA has several stages – the first of these is a clock auction, where the prices rise each round and participants indicate *how many* of the items they would like. There is then a supplementary round, where companies can make further sealed bids for package or single items. To prevent aggregation risk, participants win either one of their packages or none.

The disadvantage of the CCA is its complexity, which has two aspects: enormous numbers of combinations and confusion about second prices.

Enormous Numbers of Combinations

First, the number of possible combinations can be enormous, meaning that the process of determining a winner by choosing the optimum combination is potentially a challenge which goes close to the limits of current mathematical research.

For example, in the case study above let us assume that there are the following number of licences:

3×800 MHz, 4×2.6 GHz and $3 \times 1800 = 10$ licences in total

There are 1024 different combinations that each bidder could submit in the supplementary round, and if we assume there are five bidders that is 5120 bids in total. We would also include all the bids from the clock round, which if we assume that has 10 rounds that adds 50 to the total number of bids which now equals 5170.

In theory, there are billions of ways in which these 5170 bids could be combined. However, there are branches of maths and computing devoted to solving this kind of issue, and problems with tens of thousands of variables are frequently attempted. These types of combinatorial auctions have been run successfully, and it seems widely accepted that a successful auction could be run with the number of supplementary bids limited to say 100. In any case not all bidders may wish to make the maximum number of supplementary bids.

But it must be remembered that CCAs fundamentally rely on complex and advanced mathematics that can only be addressed by firms who can afford to have the relevant skills in-house, or who can afford to hire expensive consultants. Other auction formats, by contrast, do not present this obstacle to competition. This continues to be a concern for bidders and regulators alike.

Confusion about Second Prices

The second aspect of CCAs' complexity is the determination of the second price.

Where there are multiple items on sale, the second price is calculated by removing each of the winning bidders in turn and recalculating the total value of this new set of winning bids. This figure is then deduced from the original total, and this difference is deducted from the amount the winning bidder offered to pay.

If no one else has bid on your winning package, then the total of winning bids with you removed is likely to be much lower than the original total. This means that your second price will be much less than your original bid.

This can lead to unexpected results like in the 2012 Swiss auction where one bidder pays much less than another for similar spectrum, as shown in Figure 18.6.

In the Swiss auction, both Orange and Sunrise won 160 MHz but Sunrise paid €270 million more. Sunrise even paid €100 million more than Swisscom despite the latter winning considerably more spectrum – 255 MHz. Looking at the amount paid per MHz per head of population (€/MHz/pop in Figure 18.6), Sunrise paid three times as much as Orange and twice as much as Swisscom.[*]

How did this happen? One or both of the other two companies[†] put in bids for the lots that Sunrise won, as slightly below Sunrise's own price. This is known as predatory bidding and meant that Sunrise's second price was only slightly below the amount it bid.

Frequency band	Orange	Sunrise	Swisscom
800 MHz	20	20	20
900 MHz	10	30	30
1800 MHz	50	40	60
2.1 GHz FDD	40	20	60
2.1 GHz TDD	–	–	–
2.6 GHz FDD	40	50	40
2.6 GHz TDD	–	–	45
Total spectrum won (MHz)	160	160	255
Total paid in Euros	128,811,597	401,100,973	299,623,393
€/MHz/pop	0.103	0.320	0.150

Figure 18.6 Spectrum won (MHz) in the February 2012 Swiss mobile auction.

---

[*] *PolicyTracker*. (2012). Incumbents triumph in Swiss multiband auction. https://www.policytracker.com/headlines/incumbents-triumph-in-swiss-multiband-auction

[†] The bids were not revealed after the auction so we cannot know who was responsible for the predatory bidding.

However, Sunrise did not engage in predatory bidding, saying that it did not have the funds to meet these obligations if their calculations went wrong and the predatory bids turned out to be winning bids. This meant that their competitors' second prices were dramatically lower than their first prices.

The outcome caused much concern among companies planning to bid in a CCA held in Ireland the following year. One company, Telefonica* said the format may 'have failed to give bidders clarity and certainty as to the level of expenditure that they were liable for as a result of the bids that they placed'. Another operator, Meteor, said, 'a similarly asymmetrical outcome in Ireland would be highly damaging to the competitive functioning of the market and contrary to [the Irish regulator] ComReg's statutory objectives'. Sunrise's chief executive Oliver Steil said the company experienced the results of a 'very unfavourable auction format' that benefitted 'companies with a strong balance sheet'.[†]

'The CCA is not without its flaws. It solves the aggregation problem, but at the same time appears to offer the opportunity to "punish" winners if their package is easily identified', said Dennis Ward, former auctioneer for Industry Canada. 'In these instances, some winners may be saddled with financial obligations that may diminish their competitiveness in the post-auction market'.

There have been further CCAs since 2012 without the same price disparity, but the Swiss result certainly gave the reputation of auctions a serious knock. It undermined the feeling that auctions could be refined and enhanced to resolve any problem which might be thrown at them. The Swiss auction showed that the increasing complication could have disastrous consequences, rather like using a bullwhip to swat a fly on a friend's face – a very risky enterprise.

---

* See *GSM Liberalisation Project: Publication of non-confidential submissions to Document 11/75 and correspondence provided by respondents* published by Irish Commission for Communications Regulation 15.3.12, available at http://www.comreg.ie/publications/gsm_liberalisation_project_publication_of_non-confidential_submissions_to_document_11_75_and_correspondence_provided_by_respondents_and_comreg_written_responses_to_same.583.104049.p.html

† Taken from slides released at Sunrise investor call 24 May 2012.

Drawing Together the Strands

It is useful to briefly summarise the perceived flaws of auctions:

- The default mechanism of auctioning spectrum has contributed to the creation of an oligopoly instead of a vibrant, competitive mobile market.
- High auction prices prevent the speedy roll-out of new technologies and services.
- Auctions are effective in new markets but more problematic when the market becomes established.
- Incumbents can use auctions to obstruct competition.
- Governments have become dependent on revenues from spectrum auctions, creating a new form of 'regulatory capture'.
- The constraints placed on auction bidders, such as caps, floors, market share limitations and coverage obligations, can obviate the purpose of the auction, namely, to assign the spectrum to the highest bidder.
- Complex auction formats like the CCA can damage competition through differential pricing.

There are certainly some powerful arguments here, but one wonders whether auctions are being made to shoulder too much blame for wider problems in the mobile sector. For example, the mobile market is also concentrated in countries which have not assigned licences via auctions, such as Bahrain, Egypt and China, which all have three operators.

Concentration is also a common feature in many mature industries where access to a scarce resource such as spectrum does not create a barrier to entry. In the United States, the search engine market and the arcade, food and entertainment complexes market are both more concentrated than the mobile market, according to a survey in 2012.*

And have not the bidder constraints, criticised above, alleviated some of the negative tendencies of auctions? For example, the reservation of spectrum for a new entrant in the 2012 Dutch auction increased the competitiveness of the national mobile market

---

* IBISworld 10.2.12 Top 10 Highly Concentrated Industries http://news.cision.com/ ibisworld/r/top-10-highly-concentrated-industries,c9219248

by giving an 800 MHz slot to a smaller player: the MVNO and 2.6 GHz licence holder, Tele2.

The auction critics recommend tackling the mobile oligopoly by introducing LSA and by auctioning more bands, but they do not say how you should assign major mobile bands where the big incumbents are likely to be the only participants. Zehle predicts a return to administrative assignment but views this option with concern. Are regulators more likely to make pricing mistakes than the market? Surely their market knowledge will be worse than that of the operators?

You may have noticed that when we examine the criticisms of auctions, we tend to return to the same arguments made in the 1990s when auctions were replacing beauty contests or fixed prices. However, the growing concern about auctions does point to the weaknesses of the liberalisation project and how auctions might contribute to those problems.

The reputation of market-based assignment methods has certainly taken a dent in recent years but these criticisms have yet to land a killer blow. Many more hits will be required to wrench the auction treasure chest from the grip of government ministers.

Perhaps, the lasting contribution of the more recent criticisms of auctions will be a recognition that a variety of approaches are needed if you view wireless communications as an oligopoly which needs to be opened to further competition. Auctions *per se* have not had a substantial impact on competition: perhaps sharing, LSA and whitespace will be more successful. Strict adherents of Coase would argue that *all* spectrum should be auctioned: maybe experience is teaching us that a combination of assignment methods – auctions, unlicensed, administrative and sharing – produces the best use of the airwaves.

PART IV
THE NEW AGENDA

19

Introduction to Part IV
Finding the
Right Metaphor

In this section, we consider the new approaches to spectrum management which are attracting attention following a growing realisation of the limitations of the liberalisation agenda. It is tempting to say that the liberalisation paradigm is over and we are entering a new one, but this metaphor seems inadequate.[*]

The phrase 'paradigm shift' was originally coined by Thomas Kuhn in the early 1960s[†] and had a specific meaning. He argued that the history of the natural sciences was not characterised by the steady march of progress: it was revolutionary rather than evolutionary. Just like in the arts or literature, there were periods of rapid change when existing thought systems collapsed under the weight of anomalies and unanswered questions. From this crisis would emerge a new paradigm and the old one would be discarded. In time the weaknesses of this new paradigm would be exposed and the process would begin again.

In popular usage, 'paradigm shift' has come to mean something less specific, namely, a fundamental change in approach. But even this popular usage still implies that the previous approach has been

---

[*] For example, a very interesting collection of essays in the *Communications and Strategies* journal (No. 90 Q2 2013) was titled, 'The Radio Spectrum: A Shift in Paradigms'. This did not discuss the 'paradigms' metaphor in detail and seemed in fact to be pointing towards a more evolutionary interpretation of developments in spectrum policy. It talked of entering a 'third phase, characterized by a far richer ranger of possibilities. The prospect of applying market mechanisms across the board across the radio spectrum in the foreseeable future has vanished. [...] Policy options are increasingly granular, increasingly differentiated, increasingly sophisticated. Shared or collective use of the radio spectrum is increasingly both a focus of the technology and policy environments' (p. 12).

[†] Kuhn, T. (1962). *The Structure of Scientific Revolutions*, University of Chicago Press.

abandoned, and this is a serious weakness for our discussion of spectrum management.

It may be useful to consider the development of purely academic subjects in terms of paradigm shifts, but it is problematic when applied to policy making. Spectrum management is influenced by the academy, but it is a set of laws, regulations and policies which govern behaviour in the practical world. Policies do change but it is rare for previous approaches to be abandoned entirely.

Has spectrum liberalisation been abandoned? Certainly not. Technology-neutral licensing and spectrum auctions are normal practice in the majority of countries; trading and spectrum pricing are commonplace in a significant minority. The policies associated with spectrum liberalisation are continuing, although their limitations are evident.

A Mixture of Tools

The humble toolbox – often referred to in discussions of regulation - is a more appropriate, if prosaic, metaphor. In a carpenter's toolbox, we can find implements thousands of years old. Nails, for example, were being hammered at least 5000 years ago in Ancient Egypt, but most house builders now use a nail gun instead. However, the hammer is still the best tool for more delicate or smaller jobs and no modern toolbox is complete without one.

The policies associated with spectrum liberalisation will remain part of the regulatory toolbox for many years to come, just as command and control continues to be the dominant mode of assigning spectrum in some sectors, such as military usage. The virtue of the toolbox metaphor is that it focuses attention on a central puzzle in spectrum management: the overall impact of applying different policies in different sectors.

The 'paradigms' approach also does not recognise that the current focus for policy development – sharing – was part of the liberalisation agenda. Market mechanisms work so successfully in land because they encourage sharing. A piece of agricultural land may have a freehold owner who rents it out to several leaseholders for the purpose of growing crops, but other farmers may have bought a right to graze their animals on the land at certain times; there may be water abstraction rights on the land, hunting rights and even fishing rights if a river runs through the field.

Liberalisers wanted to apply a similarly wide range of property rights to the spectrum; hence, their support for underlay technologies like ultra wide band, overlays like TV whitespace, and giving spectrum licensees the ability to lease out their spectrum. What we have seen in recent years is new interest in the sharing aspect of the liberalisation agenda.

In this final section, we focus on these new approaches to sharing: Licensed Shared Access (LSA) as a development of the leasing model and the wider concept behind TV whitespace: Dynamic Spectrum Access. We also consider the likely impact on spectrum policy of 5G and other new technologies.

20
LICENSED SHARED ACCESS

Introduction

Not all of the spectrum that has been identified globally for International Mobile Telecommunications (IMT – The ITU's official name for mobile communication services) is free from existing uses. In some cases, it is a prerequisite that the existing use is ended before a new use can begin, but in others the spectrum may be held by organisations that could be in a position to allow IMT usage but have no specific regulatory method of doing so. Licensed Shared Access (LSA) has been proposed as a method to permit these users to release some of their spectrum for other services.

In this chapter, we explore the issues with gaining access to spectrum that has been identified for IMT but which may be in daily use by other services, and the proposals for a scheme (LSA) that would enable these users to share their spectrum.

Availability of IMT Spectrum

Globally, there is upwards of 1150 MHz of spectrum that is identified by the International Telecommunication Union (ITU) for IMT, namely,

- 450–470 MHz
- 698–790 MHz (some countries only)
- 790–960 MHz
- 1710–2025 MHz
- 2110–2200 MHz
- 2300–2400 MHz
- 2500–2690 MHz
- 3400–3600 MHz (over 80 administrations)

Within these bands, there are specific frequency ranges that have been harmonised at a regional level for use by mobile services – that is to say that a regional body responsible for setting standards has agreed to a particular pattern of use. Typically, the harmonisation of bands is along ITU region lines and differs from region to region; however, some bands are used in the same way in all three regions (they are globally harmonised). Globally harmonised bands are particularly popular as they are more likely to be built into every device, and the larger ecosystem would offer bigger economies of scale and lower equipment costs.

Most of the bands for IMT services use paired spectrum – that is to say that one piece of spectrum is used for the uplink (the connection from the mobile terminal to the base station) and another piece is used for the downlink (the connection from the base station to the mobile terminal). This paired or duplex spectrum requires that a gap be left between the uplink and the downlink pieces in order to avoid the problem of having a high-powered base station transmitter on a frequency immediately adjacent to weak signals from mobile terminals that it is trying to receive. Some bands have two arrangements, one in a paired format and another in an unpaired format, and it is down to each administration to decide which to adopt. Significant problems can occur along borders of neighbouring administrations that take different approaches (i.e. where one country chooses a paired format and another chooses an unpaired format) as this leads to base stations on one side of a border transmitting directly on frequencies used by base stations on the opposite side of the border to receive the weak signals from mobile devices. Practically speaking, due to these centre gaps and the different regional uses of the spectrum, there is typically only around 1000 MHz of the ITU identified spectrum that is actually usable in each region for IMT services (Figure 20.1).

Note that the amount of spectrum harmonised at a regional level changes from time to time, but at the time of writing it constituted

ITU region	Typical amount of regionally harmonised IMT spectrum
1 (Europe, Middle East, Africa)	825 MHz
2 (The Americas)	540 MHz
3 (Asia, Pacific)	655 MHz

Figure 20.1 Typical amounts of IMT spectrum harmonised in the three ITU regions.

only about 60 per cent of the spectrum identified and usable for IMT. The remainder is yet to have a common pattern of use identified, though in some cases approaches used in one region may be imported into others. This means, for example, that in some parts of the Americas, bands that are harmonised in Europe and Asia may be adopted even though they do not match the regionally harmonised approach. In Argentina and Venezuela, for example, mobile services exist in the European/Asian 900 MHz band, in addition to the regionally harmonised American 850 MHz band.

The identification of IMT spectrum at the ITU level does not mandate administrations to set it aside for this purpose, and there continues to be extensive use of some of this spectrum by existing services. For example, in many countries, the band 2300–2400 MHz is often used for government or military services, or services ancillary to broadcasting (SAB) (e.g. wireless video cameras). Similarly, the band 3400–3600 MHz is part of the spectrum used by C-Band satellite downlinks, which are heavily used in regions of high rainfall where higher frequency satellite bands are of less utility due to the attenuation caused by the rain. Where spectrum identified for IMT services is being used by other services, the mobile community has been keen in identifying a mechanism by which they could gain access to the spectrum without forcing the incumbent user to relinquish his or her rights to use the spectrum or move out of the band.

The method proposed for gaining access to spectrum identified for IMT but in use by other services is known variously as Licensed Shared Access or Authorised Shared Access (ASA). Though the term *LSA* is now in most common usage, we will use the terms 'authorised' user and 'incumbent' user to define the user-granted access through an LSA approach and the user who is originally assigned the spectrum, respectively.

How Does Licensed Shared Access Work?

The concept of LSA is that the incumbent spectrum user authorises one or more other users to share its spectrum under a defined commercial and regulatory framework. This differs from spectrum trading insofar as the incumbent user has usually not gained access to their spectrum through a commercial award procedure (e.g. an auction) and may not be able to trade it. Further more, the incumbent user retains primary use of their spectrum. In many countries, for example, government users have no legal means of trading the spectrum they have been assigned. LSA is totally voluntary and incumbent users are not mandated to offer shared access. The authorised user may need to operate on a non-interference basis (i.e. the user must cause no interference to the incumbent user); however, it does require the incumbent user to leave the spectrum free for the authorised use.

The European Commission's Radio Spectrum Policy Group (RSPG) defines LSA as follows:

> A regulatory approach aiming to facilitate the introduction of radio-communication systems operated by a limited number of licensees under an individual licensing regime in a frequency band already assigned or expected to be assigned to one or more incumbent users. Under the Licensed Shared Access (LSA) approach, the additional users are authorised to use the spectrum (or part of the spectrum) in accordance with sharing rules included in their rights of use of spectrum, thereby allowing all the authorised users, including incumbents, to provide a certain Quality of Service (QoS).*

To take an example, the military in a particular country may have been allocated the band 2300–2400 MHz long before it was identified for use for IMT. Through re-planning or re-engineering of their services, they may be able to permit access to other users to some or all of this band in some areas or across the whole country. Rather than return the spectrum to the regulator to be licensed through a commercial process, the military may alternatively offer one or more other users the potential to share the band by agreeing to LSA. The LSA may be of a fixed duration and may contain a number of restrictions on usage to protect the military user. It may also contain commercial terms, generating revenue for the military incumbent.

There are various ways in which LSA could be implemented:

- It may be used as a means to allow early access to a band that is in the process of being re-farmed for long-term IMT assignments, as a transitional means to gain access to the spectrum

---

* Radio Spectrum Policy Group Opinion on Licensed Shared Access, European Commission, November 2013.

until such time as the re-farming is completed and the band can be assigned through a typical commercial award procedure.

- It might be used to allow the authorised user to share the band with the incumbent user over an extended period, but in a situation where the incumbent user cannot fully re-farm the spectrum, or in which their future use will continue and may change.
- It may permit access to small gaps (in geography or spectrum) that would not, otherwise, form a sufficiently contiguous patchwork to make it worthwhile assigning spectrum on a more definitive basis.
- It may also be a way of opening up a particular band for cognitive radios or using a database-driven approach to identify specific frequencies and locations where the authorised user may operate without causing interference to the incumbent.

In principle, LSA could apply to use of the incumbent's spectrum for purposes other than IMT: private mobile networks could be established within a government band. Alternatively, SAB use of spectrum could be enabled through LSA. It is, however, the potential to find more spectrum for IMT that is driving the concept of LSA forward.

The Appeal of LSA

While LSA may seem to some to be just a back-door method of transferring usage from one user to another – similar to spectrum trading – unlike trading it does not need authorisation by the appropriate national administration, nor does it require the incumbent user to be in a position to trade spectrum. It may also be quicker than a full re-farming and re-assignment process as it does not require the incumbent to completely re-organise their usage to make way for the authorised user. Further more, it gives the incumbent user continued rights to use the spectrum at some future date, should they wish to retain access either through a fixed-duration contract, or through clauses that allow the incumbent to preempt usage by the authorised user (though such a clause would not be commercially popular).

Perhaps more tellingly, it may also permit commercial mobile operators to gain access to spectrum more quickly and at a lower cost than through traditional award procedures. The lower cost stems from the inherently less attractive nature of spectrum that is not exclusively owned by the authorised user. It may also provide a means of getting access to spectrum that would otherwise be unavailable if standard processes and procedures concerning spectrum assignment had to be followed.

A study conducted for Ericsson, NSN and Qualcomm by Plum Consulting* in 2013 estimated that the economic value of opening up the 2300–2400 MHz band to LSA in Europe is to be between 6.5 and 22 billion Euro, compared to the situation where the band remains in the hands of the incumbents. This assumes that this band, though used by incumbents (largely government users), is not used in its entirety and that some of the band could be released in each country. Where the band is already in full use, there would be zero benefit in LSA as the band could not be made available for an authorised user.

Conversely, access to spectrum using LSA may be on worse technical and contractual terms than through a traditional assignment process. Access may be geographically patchy, may come with stronger than usual restrictions and emission limits to protect the incumbent user. Usage may also be able to be preempted by the incumbent user at relatively short notice (i.e. on much shorter time scales than typical spectrum licences), meaning that authorised users of the spectrum may need to be ready to terminate or modify their networks in far shorter time scales than for networks operating in IMT spectrum licensed on normal terms. Thus, the quality of service that authorised users may be able to offer may be less than that in more traditionally assigned IMT spectrum.

LSA in Practice

While much effort has been put into trying to develop a harmonised regulatory framework to permit LSA, the most focus, particularly in Europe, has gone into the 2300–2400 MHz band, as the typical

---

* Lavender, T., Marks, P., and Wongsaroj, S. (2013, December). The Economic Benefits of LSA in 2.3 GHz in Europe. A report for Ericsson, NSN, and Qualcomm (http://www.plumconsulting.co.uk/pdfs/Plum_Dec2013_Economic_benefits_of_LSA_2.3_GHz_in_Europe.pdf).

ownership of that band (e.g. the military) is most conducive to an LSA approach. There are other potential bands to which LSA could apply, such as the aforementioned satellite spectrum as well as the remaining UHF television broadcasting band from 470–790 MHz; however, the battle for the use of the UHF spectrum seems to be divided firmly between those who support the continued use of the band for broadcasting and the mobile industry who see the long-term closure of terrestrial broadcasting (in favour of IP-based video delivery) as being inevitable. In this situation, LSA could be used to provide access to the UHF broadcasting band in those countries, or in those areas where usage is not ubiquitous while continuing to allow countries and regions where terrestrial broadcasting is the primary source of linear video to continue to use the band.

In this respect, as with the 2300–2400 MHz band, LSA may just be a means of providing a 'foot in the door' to using the band, to prove the viability of sharing the band. The long-term goal of any spectrum user has to be to secure exclusive use so as to be able to offer a reliable quality of service. Even though LSA may provide a means to do this on a temporary basis, it is, by its nature, never going to provide a permanent, high-quality means of providing a service. It is, instead, another way of permitting users to share spectrum on what could be argued to be a less ad hoc basis than, for example, cognitive radio or Dynamic Spectrum Access.

21

COGNITIVE RADIO AND DATABASE-DRIVEN SPECTRUM MANAGEMENT

Those looking to reform the command-and-control approach to spectrum have always been very interested in cognitive radio, which senses whether a frequency is being used and only transmits in unoccupied bands. Cognitive radio was appealing for two reasons: it means that spectrum could be used more efficiently but it could also be the foundation for a sophisticated commercial market in spectrum. In 1998, Eli Noam* explained how cognitive radio could allow many buyers and sellers to take part in real-time automated electronic auctions to gain access to frequencies, ending the structural problems with the spectrum market.

Writing in 2015, Noam's vision is still many years away. Implementing a full cognitive approach proved challenging, and attention then turned to a hybrid system which used databases rather than sensing to detect unused channels.

Cognitive Radio

The concept of cognitive radios was already briefly introduced in Chapter 11 while discussing approaches to the use of whitespace spectrum. At the most basic level, the concept is relatively straightforward: two devices which wish to communicate listen to the radio spectrum around them and if both identify that a frequency is not being used, they use it and theoretically do so without causing interference.

---

* Noam, E. (1998). Spectrum Auctions: Yesterday's Heresy, Today's Orthodoxy, Tomorrow's Anachronism. *Journal of Law and Economics*, Vol. 41, No. S2, pp. 765–790.

This methodology is actually already used by a number of radio services. Take, for example, a radio amateur. They are licensed to use specific frequency bands, but within those bands they can use any frequency that they wish (subject to some basic band planning to try and ensure that different kinds of service congregate together to make them easier to find). The radio amateur will switch on his radio and tune around the band which he is interested in using, and which may be in use by many other stations, to find an empty frequency. Given the normal reciprocity of radio propagation (i.e if I cannot hear you, you cannot hear me), it stands to reason that an apparently empty frequency should be clear. For radio amateurs, it is normal practice to first check that the frequency is actually clear by announcing one's intention to use it. If this initial announcement is not met with any response, the amateur will then continue to operate on that frequency.

This process generally works very well and provides a means for stations to share the same frequency band without the need for a central co-ordinatory function to assign specific frequencies to specific stations. The ethos of cognitive radio is similar, just that the processes are automated in software.

There are, however, several critical differences between the amateur radio example, and that of a cognitive radio device. In particular, the first proposed uses of cognitive radio were for them to operate inside a television broadcasting band. Part of the logic behind this is that the television broadcasting band is a relatively stable environment, where frequency use changes little from day to day and thus there is no need to regularly check for other users. Despite this, using a cognitive radio in a television band still raises a number of problems.

Difficulties in Avoiding Interference

First, the cognitive devices could be at ground level, or indoors, but the main receiving antenna where interference could be caused by transmissions from the cognitive device could be at roof level. This would make it very difficult for the cognitive device to ascertain whether or not a frequency is being used as the signal at ground level may be far below that on the rooftop and thus essentially undetectable.

While cognitive radios can be made very sensitive to particular signals (and thus be able to detect and avoid even very weak signals), any change in that signal (e.g. an upgrade in broadcast technology) could make those detection algorithms ineffective.

Conversely, if the cognitive device identifies a weak signal, it does not know whether this is being used locally. Its only option is therefore to avoid using the frequency, whereas in reality it may be perfectly feasible to use if it is too weak to provide a service and is therefore not in use, ruling out the use of frequencies that would otherwise be free.

Television receivers have no way of communicating with the cognitive device if they experience interference. While the cognitive device could conduct a first, tentative transmission to allow interference to be tested (in the same way that the radio amateur does), there is no way for any nearby television receivers to respond.

Even if the television receiver could communicate with the cognitive device, it may not be clear which particular device was causing the problem, as the area over which the television may experience interference may be quite large (though this could be overcome by allowing the cognitive device to send an identifier along with any test transmission), or the problem may be caused by interference from multiple cognitive devices simultaneously transmitting.

First Trials

Despite these problems, a number of major corporations became interested in the concept and developed prototype devices. The FCC conducted tests of these devices in July 2007[*] and all of those submitted for testing were able to detect television signals on an occupied channel. However, the concept of the use of a cognitive radio to gain access to television whitespaces continued to concern the incumbent spectrum users (e.g. the broadcasters). Two specific concerns dominated:

- First, the FCC's tests had been designed to detect a specific standard of broadcast technology. Broadcasters were concerned that if this technology was changed at some future date, the devices may not work successfully for a new kind of signal.

---

[*] https://apps.fcc.gov/edocs_public/attachmatch/DA-08-2243A2.pdf

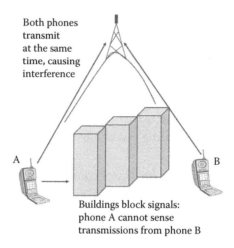

Figure 21.1 The hidden transmitter problem.

- Other users were similarly concerned that even an update to the technology they used may render the cognitive devices unable to detect their usage.

In addition, in the case where a cognitive device is sharing spectrum with, for example, a mobile radio user (instead of television transmitters), there is the 'hidden transmitter' problem (Figure 21.1). The cognitive radio could be behind an obstruction that shields it from being able to detect a transmission from another user. Believing the channel to be free, it may then transmit on the apparently empty frequency yet by doing so cause interference to nearby receivers whose reception was not blocked by the same obstruction.

Cognitive Techniques Already in Use

Cognitive principles, however, can be, and are, applied to other technologies to achieve similar results. For example, the 5 GHz band used by many Wi-Fi hotspots is shared with radar services. The use of the band is constrained in many countries by the need to implement 'detect-and-avoid' techniques. Essentially, if the hotspot detects a radar on a frequency it is using, it must change frequency to leave the radar free to operate. This technique, known as Dynamic Frequency Selection, shares many of the same principles with a cognitive radio but many of the same problems, too (e.g. indoor signals versus outdoor and the inability of the radar to inform the hotspot if it is suffering interference).

Probably the nearest system in use today which operates along cognitive lines is the cordless telephone technology DECT. In Europe, there are 10 specific channels identified for use by DECT equipment. Each DECT base unit will examine those channels and identify which is the clearest of interference. When required to connect to the handset, it then uses the channel it deems least interfered. This process is known as Instant Dynamic Channel Selection and according to the ETSI guide to DECT*, it 'avoids any need for traditional frequency planning and greatly simplifies the installations'.

There are a range of other technologies which employ a detect-and-avoid principle including Bluetooth and even military systems for whom avoiding interference is second nature insofar as their need to dodge jamming signals from the enemy. In each case, the idea of reusing spectrum where it is available (e.g. whitespace) while trying to avoid causing interference to the incumbent user is employed to try and share and make better use of the frequencies concerned.

Cognitive Devices in Public Sector Spectrum

Though the concept of cognitive radios operating in the television broadcast bands has not, so far, received full regulatory approval, proponents of cognitive devices continue to explore options for their use. Many have eyed public sector spectrum, in particular that used by the military, as a potential playground for cognitive devices. They argue the following:

- Away from military bases, the use of military spectrum is scant.
- In most other areas, usage tends to be around military exercises which are infrequent, tend to be geographically focussed, and for which the relevant information (e.g. date and time) are known in advance such that they could be avoided.

---

* *ETSI. Digital Enhanced Cordless Telecommunications (DECT): A High Level Guide to the DECT Standardization*, ETSI TR 101 178 V1.5.1 (2005–02), Technical Report (http://www.etsi.org/deliver/etsi_tr/101100_101199/101178/01.05.01_60/ tr_101178v010501p.pdf).

- Military technology is designed to be able to deal with interference (electronic countermeasures) and thus low levels of interference ought to be able to be tolerated.
- Cognitive devices could be designed to detect military use and 'sense and avoid' them.
- The military (and other governmental users) are far too cautious and could easily share their bands with no problems.

There are precedents. It is not uncommon for programme-making equipment, in particular wireless cameras, to be permitted to share military spectrum. Such use can, however, be steered away from any particularly sensitive areas (e.g. army bases) and the well-defined, well-controlled, sporadic and occasional nature of use pose less of an issue to military users than would a 24-hour, 7-day mobile service or thousands of Wi-Fi hotspots occupying their bands.

While a move from the relatively benign environment of television broadcasting to the military arena may seem an odd direction for cognitive proponents to take, a 2009 opinion on the use of public sector (a.k.a. government) spectrum by the European Commission's Radio Spectrum Policy Group* said, 'Cognitive radio (CR) is a useful concept, which when implemented could achieve more efficient use of spectrum and should be exploited further'.

As of today, there is no real cognitive-based sharing of public sector spectrum; however, some taboos are being challenged. In the United Kingdom, Ofcom is currently examining the possibility of sharing aeronautical navigation bands with wireless microphones. Historically, any moves to share aeronautical spectrum with any other users have been met by a barrage of safety-of-life arguments and a set of sharing criteria so severely restrictive that even blowing your nose in the vicinity of one of their bands would have exceeded the proposed interference thresholds.

If terms and conditions can be drawn up for wireless microphones to share with aeronautical navigation, it could open the door to sharing with other technologies and the cognitive approach would be an obvious contender.

---

* Best Practices Regarding the Use of Spectrum by Some Public Sectors, RSPG, February 2009.

Databases

One proposed solution to some of the problems of cognitive radios involves placing the listening part of the cognitive devices in hilltop-type locations where it should be possible to detect all usage and then relay the information on available frequencies using 'cognitive beacons'. This has the advantage of allowing changes in technology to be dealt with at the beacon sites instead of in every cognitive device and hopefully overcoming the hidden transmitter problem by placing the sensing device in a location where it could see over a much wider area. The difficulty with this approach is in deciding who should operate the beacons and indeed who should pay for them.

The cognitive beacon approach has not gained much traction, and the solution which has garnered the most all-round support came in the form of Dynamic Spectrum Access (DSA, sometimes also known as Dynamic Spectrum Assignment). Instead of devices listening to the spectrum and detecting unused frequencies, they would instead identify their location (e.g. using GPS) and then query a central database which stored details of all planned use by the incumbent. The database would model the coverage of the existing transmitters and respond with a list of frequencies available at the DSA device's location. The DSA device could then register its use for the required period of time in the database, so that other devices could not then usurp its frequency.

Thus, control of the use of the spectrum would vest with the database provider who in turn can be supplied information by the incumbent spectrum user (e.g. the broadcaster). Changes in the use of the spectrum, whether a new technology or some re-planning of frequencies (or indeed a forthcoming military exercise), could be loaded into the database such that the incumbent user's spectrum would be continuously protected.

The DSA approach has grown out of cognitive radio but is based on some quite different precepts. It is not a perfect solution in that it requires the DSA device to be able to communicate with the central database before such time as it is aware of which frequencies can be used. Such communication would have to be through some other communication medium such as an existing mobile connection, Wi-Fi or even a fixed broadband line. It does, however, provide the incumbent spectrum user with much more control over how the DSA device can operate. It can control issues such as the following:

- Which frequencies are available?
- In what area they are available?
- At what time(s) they are available?
- What are the restrictions on the technical characteristics of the DSA device's transmissions?
- How do we (potentially) determine the price for access to the spectrum?

As mentioned previously, any opportunistic access to seemingly unused radio spectrum will depend to a large extent on the density of existing use, and it could be envisaged that as more of the spectrum is filled with services, the number of pockets of usable whitespace spectrum may dwindle to naught.

Ending Before It Has Begun?

Has cognitive radio had its day? In general, there is a move away from the original cognitive concept towards the database-driven, DSA approach. Substantial progress has been made in creating a regulatory framework which will allow the deployment of database-driven whitespace solutions, but 'true' cognitive radio still seems a distant prospect. Even the limited progress made with database solutions has been a slow process, with an effort lasting nearly 10 years yet to produce significant commercial deployments.

Notwithstanding this, the table may be turning on the use of military spectrum. Instead of having cognitive devices occupy unused parts of military bands, it may be more logical for the military to cognitively occupy parts of other people's bands (which is in essence what some military systems such as Link 16 already do). The ability to use otherwise empty frequencies to communicate, the fact that frequency usage is not on an assigned (and thus an easily traceable) frequency, and the ability to change frequency if the one being used becomes occupied (or jammed) would surely be good traits for a military communication system.

22

FUTURE TECHNOLOGIES

How the airwaves are managed depends partly on the characteristics of the devices being used, and in this chapter we examine upcoming technologies to consider which of their features have implications for the future of spectrum management.

For example, there is little direct regulatory intervention in the unlicensed bands where Wi-Fi operates because these are restricted to low-power devices where signals do not travel far, so there is little danger of interference. However, there are many other services where interference does not 'solve itself'. Cell phones, satellite services and broadcasting are all configured to work in dedicated spectrum which is free from other services that could cause interference. Arranging these services into bands which do not interfere with each other is a huge regulatory task, both nationally and at the International Telecommunication Union (ITU) level. It is based on a knowledge of which services use which bands.

But what if mobile were to start using different frequencies? How would this change spectrum management in the future? This chapter focusses on the main engine for technological change in the mobile arena: a new generation of services labelled 5G which are expected to use much higher frequencies.

The *way* in which the airwaves are used is also important in considering how spectrum management will need to change in the future. One new way of using the spectrum is the dynamic access methods discussed in the previous chapter. Spectrum sharing is another change to familiar access regimes and one that is expected to have increasing importance in 5G. In this chapter, we consider the implications for future policy.

This chapter also examines the possibility of using frequencies much higher than those normally referred to as the radio spectrum, including the use of visible light for data transmission.

5G

At the time of writing, 5G has no agreed definition and several industry players are working towards their own visions of the technology.

For the time being, Professor Rahim Tafazolli, who leads a research group in 5G at the University of Surrey has provisionally defined it as providing an 'always sufficient' connection that gives users the perception of 'infinite capacity'.

It seems likely that the final requirements could involve a bit rate of 1–10 gigabits per second and latency that is under 1 ms. This would enable a 'tactile Internet', where the Internet will have lower latency times than human senses, which would give the impression that the Internet is immediately reacting to an individual. According to a recent ITU-R report on the subject,

> The Tactile Internet will enable haptic interaction* with visual feedback, with technical systems supporting not just audiovisual interaction, but also that involving robotic systems to be controlled with an imperceptible time-lag.

For example, a surgeon could use a glove to perform surgery on a patient despite being possibly hundreds of miles apart from each other. Aside from medical applications, this sort of network could support virtual reality, or augmented reality.

In all likelihood, real existing 5G networks will not have to support the tactile Internet at all times and all places. Indeed, some mobile operators look at the emerging Internet of Things industry and wonder if 5G might allow the telecoms industry to displace the current plethora of proprietary networks that provide for machine-to-machine (M2M) networks. Additionally, some mobile operators have expressed an interest in entering the 'critical communications' market, which refers to communications for the emergency services, and some projects of critical national importance such as electricity generation and transport networks.

These two markets have very different needs than the tactile Internet. M2M networks typically have a huge number of connected devices over a large area transmitting small amounts of data periodically.

---

* That is, interaction by touch.

Therefore, a 5G network that supports these services need not have high bit-rates at all, but does need to support high coverage, low-cost equipment and decade-long battery lives so that engineers do not need to be deployed regularly to change them. Critical communications networks need to provide almost universal coverage inside every building over an entire territory and need to be extremely reliable, but the equipment is not as constrained by cost.

What Will 5G Networks Look Like?

In general, there are two approaches to 5G networks. One of them is a 'revolutionary' system, whose networks will look very different from contemporary 4G networks, and which has at least one new air interface. Far Eastern players such as Huawei and China Mobile generally prefer this approach. By contrast, others, such as Ericsson and Alcatel-Lucent favour an 'evolutionary' approach. This would build on current 4G technologies that are standardised by 3GPP to further optimise them with techniques such as carrier aggregation.

The revolutionaries' nightmare is that one day a network that offers an LTE-Advanced service with carrier aggregation* may market itself as 5G. All the other operators would be obliged to also brand their own LTE-Advanced networks as 5G. Pressures to do this could be compounded by the South Korean government's determination to have a 5G prototype network showcased at the Winter Olympics in 2018.

Supporting all of those user requirements in one network is undoubtedly a major technical challenge. Nevertheless, Qualcomm, Ericsson and Huawei have all argued that all of 5G's possible use cases can be catered for on one network. Cisco say they support a topology of different networks. This means that different applications will use differently designed networks, but that these network architectures, including Wi-Fi, will all be integrated into one topology.

It is generally accepted that future networks are likely to be heterogeneous. This means that the legacy macro cell network will

---

* The ability to operate a mobile network over two separate bands (e.g. 1800 MHz and 2.6 GHz). This is one of the features of LTE-Advanced.

be maintained, but that separate layers of smaller cells will be able to serve the same areas using different (or even the same) frequencies.

Core network functionality is likely to be virtualised. This means that software, which could be run on a distributed cloud, will replace the hardware that operators rely on to run their network. This technology is related to software-defined networks. Both these technologies aim to optimise network efficiency.

Another technology that has been discussed in relation to 5G is 'full duplex'. Currently, spectrum is assigned to avoid uplink and downlink transmissions in adjacent frequencies because receivers on one frequency cannot work properly if there is a very strong interfering signal on the adjacent one. But a Silicon Valley start-up, Kumu Networks, as well as researchers at the University of Texas, claim they have worked out how to transmit in both directions in the same channel. This is done by injecting the inverse of the transmitted signal into the receiver, thereby cancelling it out. They liken their technology to noise cancellation headphones. This technology, if it proves to be successful, is interesting for 5G because it will substantially boost spectral efficiency. It is also very interesting to spectrum managers because it could eliminate the need to plan spectrum bands around time division duplex or frequency division duplex, making global harmonisation easier to achieve. However, we should note that other experts consider that this technology will only reduce the need for duplex gaps, rather than eliminate them entirely, thereby achieving a more marginal efficiency gain.

Massive multiple input–multiple output (MIMO) is also likely to be employed. MIMO is already an integral part of many Institute of Electrical and Electronics Engineers (IEEE) and 3GPP standards. The concept is that a device (whether infrastructure or end-user) uses multiple antennas to form directional beams – by doing so, it can rely on reflections caused by objects between the transmitter and receiver to reuse the same frequency multiple times. Massive MIMO refers to the case where, in particular, base stations have a large number of antennas able to form a large number of beams. Massive MIMO has been considered a promising technology for the development of mmWave beam forming.

Whether or not the networks will be supported by one interface is an open question. Ericsson's Vice President Jan Farjh told a recent conference that he expects a plurality of air interfaces to make up 5G, while Huawei's Dr. Peiying Zhu predicted a unified air interface that supports different waveforms and multiple access schemes.

No matter what 5G networks will eventually look like, their construction costs, maintenance costs and spectrum consumption are likely to be considerable. The pressure from policymakers to consolidate these investments into the types of single or shared networks as described in Chapter 16 is likely to grow. The European Commission has already publicly suggested that the fifth-generation of mobile communications should be the 'last generation'. The European Commission's opinion on 5G matters because it is spending €700 million on its 5G public–private partnership scheme, which aims to drive research into 5G in the hope that Europe can retain the global position in telecoms it had during the 2G revolution in the 1990s. As part of the scheme, the private sector has agreed to invest a further €3.5 billion on research and development in this area.

How Will 5G Use Spectrum?

Many players in the spectrum industry have spoken about the use of the millimetre waves for 5G. Millimetre waves strictly refer to spectrum between 30 and 300 GHz, but the term is commonly used to refer to frequencies far higher than those currently used for International Mobile Telecommunications (IMT). The term *millimetre wave* derives from the fact that wavelengths at these frequencies are measured in millimetres, and this is often shortened to mmWave.

It is widely acknowledged that mmWaves will have some role in future 5G networks. There is debate, however, over the extent to which this spectrum can be useful outside of hyperdense urban environments. Indeed, the bulk of research undertaken by China Mobile, for example, has been on frequencies below 6 GHz, partly because higher frequencies have such poor propagation characteristics. Nevertheless, mmWaves do have a lot of benefits for 5G. First, because there is so much spectrum at higher frequencies, operators could use channels as wide as 1 GHz. This could support an exceptionally large amount of data, albeit over very short distances. Second, the spectrum is currently little used, so operators could avoid paying huge amounts for it at auction. The waves are also small enough that many antennas can fit into one device. Samsung has developed a prototype phone with 32 antennas built in.

Many researchers also argue that the propagation challenges can be overcome by new technologies, such as beam forming. Advanced software could potentially control these directional beams to allow waves to bypass obstacles such as street furniture by bouncing the signal off other street furniture. The same researchers also point to the latest Wi-Fi standard, 802.11ad, which supports the new WiGig band at 60 GHz. Millions of these devices are expected to be rolled out in 2015. These researchers point out that if an IEEE standard can cope with these extremely high frequencies, then why can 3GPP standards not cope?

Professor Ted Rappaport, director of the NYU Wireless research centre, is one of these voices. In April 2014, he said:[*]

> mmWave is not this mythical crazy place where radio waves won't work, quite the contrary: our research at NYU and others presenting here have shown quite clearly that mmWave wireless will work and there are new capabilities when you get up to these high frequencies.

Specifically, Ericsson is looking at using the 15 GHz band for Massive MIMO, Samsung is looking at super wide-band hybrid beam forming at 28 GHz and Nokia is looking at super wide-band single carrier transmission and beam forming at 70 GHz. Note that the 28 GHz band is currently used by satellite services which, as we discussed in Chapter 8, are reluctant to share with terrestrial services. Which of these companies have put their money into a winning technology is a question that only time will answer. Despite the extensive research into these bands, 5G networks are likely to be largely based upon re-farmed spectrum below 6 GHz.

Policy

The 4G assignments differed very little from 3G assignments. The bands were typically assigned through an auction and were available to mobile network operators on an exclusive basis. However, the current

---

[*] *PolicyTracker.* (2014). Research suggests millimetre waves could be used for 5G. https://www.policytracker.com/headlines/researchers-say-that-milimetre-waves-could-be-used-for-the-next-generation-for-5g

consensus on 5G appears to be that it will involve spectrum sharing. This is partly because the types of spectrum that are being discussed for 5G differ so much than for 3G and 4G. Additionally, as explained throughout this section of the book, spectrum sharing technologies are developing so rapidly that new ways of accessing spectrum are becoming more feasible.

This change in mood can be seen in several places. For example, China Mobile has publicly said that it expects spectrum access will be shared among multiple operators. Additionally, a senior figure in the spectrum team in the UK civil service, Simon Towler, said that dynamic spectrum access will 'underpin' 5G. Dr. Kent Rochford,* co-director at the US National Institute of Standards and Technology, has also said that 'a key attribute of 5G will be spectrum sharing'.

Even Eduardo Esteves, from one of the biggest proponents of licensed spectrum, Qualcomm, has conceded that 5G across all spectrum bands could be shared through dynamic spectrum access or Licensed Shared Access. In 2015, the Radio Spectrum Policy Group was also preparing a report on how to assign spectrum, and it is likely to include guidance on how to licence shared spectrum for 5G services.

Forms of spectrum sharing are currently being developed in the context of a Federal Communications Commission (FCC) proposal to share 100 MHz of spectrum in the 3.5 GHz band. The scheme, known as the Citizens Broadband Radio Service, concerns the 3550–3650 MHz band. The work was provoked by a report from the President's Council of Advisors on Science and Technology in 2012. The FCC's current vision for the band is a three-tiered spectrum access system. At the top tier, the US Navy radar system would benefit from protection from interference. A second tier would be used by Priority Access Licence (PAL) holders. PALs will be acquired through competitive bidding and last for 1 year, for one census tract, and for 10 MHz. The third layer is for general authorised users, who receive no protection from interference. Interference management in general will be done by a dynamic spectrum access system that will be more advanced versions of TVWS databases.

---

* *PolicyTracker*. (2014). 5G will probably involve spectrum sharing https://www.policytracker.com/headlines/5g-will-probably-involve-spectrum-sharing-say-policy-makers-and-engineers

Qualcomm and others have expressed a preference for a two-tier system that would be akin to Licensed Shared Access (LSA)/Authorised Shared Access (ASA), but the FCC has decided to continue pursuing its three-tier approach. If successful, Commissioner Rosenworcel and others have already spoken about extending the scheme to a 5350–5470 MHz band.

Against the trend of spectrum sharing for 5G, this may not be an attractive prospect for some national regulators as the mmWave bands are unlikely to generate high auction prices. Many players in the industry are hoping that WRC-15 will agree upon an Agenda Item for WRC-18/19 that will identify spectrum for IMT above 6 GHz. An identification of mmWaves to IMT would give the global industry a legal basis with which they can assign and build-out 5G networks.

Beyond spectrum, 5G is likely to give policymakers many more headaches than just the assignment of spectrum. If small cells are to become widespread, then the pressure from operators to simplify planning procedures for new base stations will also grow. Second, the huge amounts of data involved, including* from sensors and other devices that citizens will have no direct control over, throw up many privacy issues that policymakers need to consider.

Transmitting Beyond the Radio Spectrum

Over the last 10 years, discussions about radio spectrum policy have been framed by the search for the optimal way to meet the burgeoning demand for a scarce resource. This framework could be challenged by technological developments using frequencies above 1000 GHz and even visible light (Figure 22.1).

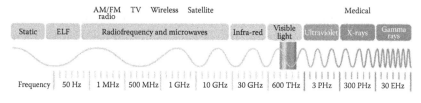

Figure 22.1 The electromagnetic spectrum. ELF = extremely low frequency.

---

* Consider, for example, readings from heat sensors or building control sensors.

Li–Fi

Visual Light Communications, which some market as Li-Fi (light-fidelity), offers one way of transmitting information without using radio spectrum. Using this concept, data can be transmitted as light from, for example, suitably designed light-emitting diode (LED) bulbs. The frequencies can still be found on the electromagnetic spectrum but within the visible (and infrared) light spectrum, rather than the radio spectrum, which is usually regarded as occupying frequencies up to 300 GHz.

Pioneers in this industry are working on a target data rate of 10 Gbps, albeit in laboratory conditions and over very short distances. Currently, developers in the field are focussing on gaining defence contracts. They argue that Li-Fi provides a high level of data security for communications over short distances. After all, as long as the curtains are drawn and the doors are closed, the data cannot escape a given room.

But ultimately, Li-Fi developers imagine consumers' handsets receiving data from the LEDs lighting their homes, or from LEDs within a widescreen television. In this case, vast amounts of data can be transmitted and received without using any radio spectrum. One could even imagine street-lights performing this task in cities at night.

Li-Fi developers point out that visible light spectrum is much more plentiful than radio spectrum, but it is also unregulated. Additionally, Li-Fi signals could penetrate water, and they claim it requires much less power to transmit signals.* While this may be theoretically true, real-world applications of the technology are still at an early stage. Further, if Li-Fi becomes an important part of how we connect to the Internet, then regulators may have to find some way of dealing with it. Because owning the licence for a certain part of the visible light spectrum seems implausible, and probably unethical, any regulation of Li-Fi is unlikely to be within the framework of spectrum liberalisation.

Tremendously High Frequencies

Another part of the radio spectrum that is currently not subject to regulation is Tremendously High Frequencies, or THz, frequencies above 1000 GHz. These frequencies are currently extremely difficult to use

---

* Not all experts agree with this claim.

for numerous practical and manufacturing reasons, but a consortium of British universities are currently experimenting with the technology. At the time of writing, the project was aiming to demonstrate communication on the frequencies within the decade. It was also focussing on using the spectrum for high-resolution imaging and understanding the effect of THz frequencies on molecular vibrations in solids. Like mmWaves, these frequencies are likely to only be used for short-distance Wi-Fi-like services.

Conclusions

Predictions about future technology are always liable to be inaccurate, and predictions about how technology will be used by consumers and to what extent it will usurp current technologies and regulation is yet more likely to be wrong. But the current discussions about 5G certainly suggest that policies about spectrum sharing and the use of higher frequencies are likely to grow in importance in coming years.

The use of spectrum outside that currently considered serviceable may also serve to change the way spectrum is valued, and we could see a new set of fierce battles between those wishing to provide mobile broadband connections and the many other incumbent users of spectrum above 6 GHz.

23
Conclusions

The driving force behind the move to a liberalised spectrum management regime has been a belief that managing the spectrum using command-and-control processes and based on technical parameters does not lead to an optimally efficient outcome. In particular, the growth in demand for mobile data and the requisite demand for spectrum has forced many to believe that these tried and tested methods of managing the spectrum could not keep up with the dynamism of the telecommunications market.

Making spectrum technology neutral, and allowing operators to update the technology they are using within their own spectrum assignments (e.g. from 2G to 3G to 4G) have undoubtedly had a beneficial effect on the ability of these operators to serve the growing demand for data.

In the introduction to this section, we argued that spectrum liberalisation should not be thought of as a paradigm which has had its day. In this conclusion, we argue that neither should it be regarded as a failure.

There have certainly been disappointments: liberalisation has not been applied to the broadcasting sector; it was not able to provide the mobile spectrum thought to be necessary to meet the capacity crunch; outside the United States there has been very little trading in high-value mobile licences and administered incentive pricing's proxy market approach has been adopted in only a handful of countries.

Similarly, in many developing countries, where demand for mobile broadband often lags behind developed countries, and where the telecommunications market is often less liberalised (and often government owned), traditional command-and-control methods of assigning spectrum continue to be more than sufficient to meet the needs of the market.

But some aspects of liberalisation have been recognised as best practice; for example, technology-neutral licensing has been adopted almost universally. Auctions have become the default assignment mechanism, and even the criticisms of them are directed at particular formats or the use of up-front fees rather than the mechanism (see Chapter 18). Spectrum trading has been a success in some countries such as the United States and in non-oligopolistic market sectors such as fixed links in the United Kingdom and France.

What has emerged from the attempts to apply a liberalised approach to the airwaves are the constraints under which any spectrum market must operate. There are the long time-lags associated with international coordination and the development of chipsets, the political pressures generated by the social importance of services such as broadcasting and defence, and the oligopolistic nature of the mobile market.

The successes of liberalisation show that this economic principle can be applied in practice to the spectrum world, but it only works to a limited extent. The new emphasis on sharing is an attempt to overcome some of those limitations, potentially offering quicker and lower cost spectrum access and avoiding some of the political and administrative issues associated with accessing public sector spectrum.

Future Prospects

To assess the long-term prospects for spectrum liberalisation, we must ask if these limitations on the spectrum market are a permanent fixture, or whether they weaken or strengthen over time. To put it another way, is it likely to get easier or more difficult to apply a market-based approach to spectrum management?

The more frequencies are sold and made tradable, as is usually the case with mobile bands, the more successful spectrum liberalisation is likely to be. One reason why liberalisation polices were unable to meet the capacity crunch was that much of the sweet-spot spectrum was being used by broadcasters or the military, and there were political reasons that prevented an open market approach to this sector.

Trends in the Broadcasting Market

In Europe, broadcasting is slowly being moved out of 700 MHz, making this a near-global mobile band (though with two incompatible band plans). In the long term, clearing TV from the 600 MHz band seems likely in Europe as well, but the Lamy report suggests this should not be considered until 2025, with implementation only possible after 2030.[*]

In the United States, TV could start vacating 600 MHz as soon as 2016 if the incentive auction process, which has been in the planning stage for several years, is a success. This is an innovative process, using a reverse auction to determine the least broadcasters would pay to vacate their frequencies. This would help set the reserve price for a 'standard'; auction,[†] where (it is assumed) mobile operators would bid to buy the bands.

So in the long term, there is likely to be a significant increase in the amount of spectrum released by broadcasters, and potentially others too, and subject to the market mechanisms associated with the mobile bands.

This process is likely to be strengthened by changes in viewer behaviour. Since the 1980s, increasing numbers of viewers have moved from terrestrial TV to cable or satellite, but since 2010 we have started watching TV on different types of devices, principally tablets and smartphones. Taking the United Kingdom as an example, Ofcom recently noted that in July 2014, 47 per cent of requests for the BBC TV catch-up service, iPlayer, were sent from tablets and mobiles, compared to just 25 per cent in October 2012 (Figure 23.1).[‡]

There is also evidence to suggest that young people watch considerably less live TV[§] than do older viewers. As shown in Figure 23.2, live TV constitutes 50 per cent of viewing for 16- to 24-year-olds but the average for all adults is 69 per cent. At this early stage in the take-up of the technology, it is hard to predict whether young peoples' behaviour will change as they get older, but a considerable change in viewing habits is certainly possible.

---

[*] See Chapter 7.

[†] There is also a process of 'repacking' bands released by broadcasters to make contiguous blocks useful to mobile operators.

[‡] Ofcom. (2014). *Infrastructure Report 2014*, p. 120.

[§] That is, watching TV at the time it is actually broadcast, rather than accessing it later using a catch-up service. The programme does not necessarily have to be broadcast in real time – such as coverage of a football match.

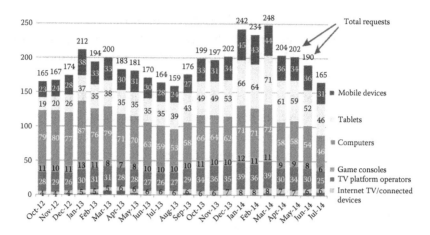

Figure 23.1   Requests made to BBC iPlayer by device type (2012–2014). (From *BBC and Ofcom 2014 Infrastructure Report*. With permission.)

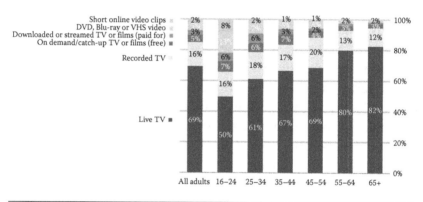

Figure 23.2   TV viewing habits by age group. (From *Ofcom 2014 Infrastructure Report*. With permission.)

Both these trends would increase the use of TV delivered through the Internet (IPTV)* while eroding the use of other platforms such as digital terrestrial television. They are being driven by increasing broadband penetration, which is also a policy goal for most governments.

The rise of Long Term Evolution (LTE) broadcast should also be mentioned here. This allows some LTE network capacity to be used for one-to-many services such as the broadcasting of video or audio signals. These can be picked up on a standard LTE phone (or tablet),

---

* Internet Protocol Television – that is, TV delivered over the internet.

without the need for an additional chip, as was required for the failed DVB-H mobile TV service.

Writing in early 2015, LTE broadcast had been available for a year or so, and initially its main use was expected to be the viewing of video clips in sports or entertainment venues. However, it could be used to offload video traffic from LTE networks by encouraging subscribers to watch a restricted menu of simulcast programming. Whether this will take off is debatable: mobile TV predecessors MediaFLO and DVB-H were unsuccessful, but the technological possibility is there which could further undermine terrestrial television.

The End Game

Cognitive radio was often seen as the ultimate goal for spectrum policy. If receivers could scan a wide range of channels for transmissions, and only use a frequency when they had sensed the channel was clear, we would have a functioning spectrum market, as Eli Noam explained in his famous 1998 paper.[*]

Noam argues that spectrum, which was initially controlled by the state, will eventually become an open market:

> It took... almost fifty years to see [the] auction paradigm implemented. Similarly, the proposed open access paradigm is not likely to be accepted anytime soon. But its time will surely come, and fully bring the invisible hand to the invisible resource.[†]

The problem, as Noam acknowledges, is time scale. His open market in spectrum relies on affordable, reliable and pocket-sized cognitive devices, and they do not seem to be on the horizon. Cognitive radio, as the saying goes, has been 'just around the corner' for the past two decades and looks likely to stay that way for many years to come other than in the somewhat limited means offered by dynamic spectrum access (DSA).

It is very hard not to share Noam's vision as the ultimate goal for the most effective and efficient use of the airwaves. Operators could

---

[*] Noam, E. (1998). Spectrum Auctions: Yesterday's Heresy, Today's Orthodoxy, Tomorrow's Anachronism. *Journal of Law and Economics*, Vol. 41, No. S2.

[†] Ibid., p. 786. The 'invisible hand' is how the famous eighteenth-century economist Adam Smith described markets.

buy spare capacity almost minute-by-minute in a real-time spot market from a variety of wholesalers, both private and governmental. Together with software radios, able to operate on any frequency, the need to operate in a specific band would be much reduced or eliminated, so removing two of the biggest barriers to a functioning market: international coordination and harmonisation. The desire of forward-looking companies such as Google and Microsoft to develop the use of TV whitespace and operate spectrum databases suggest this 'open-access spectrum' approach is not just a pipe dream.

Here we get a sense of the overall direction of travel for spectrum policy. It may be decades away, but cognitive radio and its less-sophisticated cousin, database-driven spectrum access, can remove the barriers which have held back the attempts to implement spectrum markets described in this book. In developed markets, it is very hard to imagine a return to civil servants assigning spectrum. The new emphasis on sharing will make access to the spectrum market easier, and the gradual removal of TV from the UHF band, and other incumbent users from highly valued spectrum, will make more of the airwaves subject to market mechanisms. In the much longer term, progress towards DSA will further erode the barriers to liberalisation.

In the coming years policymakers will no longer be asking "can we apply market mechanisms to the airwaves", but "what is the reason for treating this valuable resource differently from any other commodity?" Over time, that justification is going to get weaker and weaker.

Appendix A: What Does the ITU Do and How Does It Work?

The International Telecommunications Union (ITU) is the United Nations (UN) specialised agency for information and communication technologies (ICTs). Surprisingly it was formed in 1865, substantially predating the UN. The ITU was originally known as the International Telegraph Union and was set up because telegraph signals sent from one country to another had to be stopped and translated into the technical system used in the neighbouring country. The aim was to develop common international technical standards and that seamless interconnection remains a key part of the ITU's job, though their brief now includes a wide range of ICTs. The ITU has three departments, and the section dealing with standardisation is known as ITU-T.

The International Telegraph Union changed its name in 1934 and became a specialised agency of the UN upon the latter's founding in 1947. It is the only part of the UN to be funded by membership and in addition to its 193 member countries, it has over 700 members from the private sector and academia. One of the ITU's goals is improving access to ICTs for underserved communities worldwide, and this is handled by the ITU-D section.

However, it is the third part of the ITU's mission which interests us most: allocating global radio spectrum and satellite orbits. The department responsible for this is called ITU-R.

Information about global spectrum allocations is contained in a document known as the Radio Regulations. This states which services are entitled to protection from interference in any of the ITU's three global regions. ITU Region 1 includes Europe, Africa and the Middle East; Region 2 incorporates all of the Americas and Region 3 includes all of South and East Asia, as well as Oceania. (See Chapter 16, Figure 16.1.)

In its day-to-day work, ITU-R produces a wide range of reports, studies and recommendations which inform discussions about amendments to the Radio Regulations.

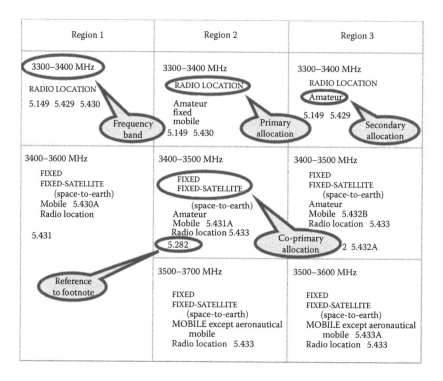

If a Radio Regulations entry is given in BLOCK CAPITALS, then that service has a primary allocation in that band. This means that these services have priority in any given frequency. If the entry is in lowercase, then a service has a secondary allocation. This means they may use the band but cannot expect any protection from interference from the primary allocation, and must not interfere with primary use of the band. Confusingly, some bands have several services in BLOCK CAPITALS. When this occurs, it is called a co-primary allocation. In these cases, both services have equal priority in a band, and it is in effect up to the nation state to assign the services to one band if both services cannot share the spectrum. To confuse things further, nation states can secure footnotes if they wish to be exempt from the regulations, or if they wish to add a caveat.

A spectrum band is *allocated* to a service by the ITU-R in the Radio Regulations, whereas nation states *assign* a licence to use all or part of a band to a particular user, such as a broadcaster or mobile operator.

The World Radio Conference Process

Unlike national assignments, the Radio Regulations can be amended only once every 3 or 4 years at World Radio Conferences (WRCs). Preparations for these conferences take years, and the conferences are often gruelling and highly politicised processes. The Agenda Item for each conference is set at its predecessor. WRC-12 set WRC-15 with 44 Agenda Items.

Preparation is done on two levels. Technical preparations are made by the industrial Working Parties inside the ITU-R's various study groups while common positions are prepared by five regional organisations. Study Group 4 deals with satellite services, SG 5 deals with terrestrial services, SG 6 deliberates on broadcasting and SG 7 examines the use of spectrum for science.

Of the regional organisations, the Inter-American Telecommunication Commission (CITEL) represents the Americas, Arab Spectrum Management Group (ASMG) represents the Arab states, the African Telecommunications Union (ATU) represents Africa, European Conference of Postal and Telecommunications Administrations (CEPT) represents Europe (including Russia) and the Regional Commonwealth in the field of Communications (RCC) represents the former Soviet Union. These groups will prepare common positions and common briefs. These regional bodies have no formal power, but they allow regional harmonisation and assist the formation of consensus, which is always preferred by ITU-R over voting. If votes do occur, then each administration has one vote: the biggest country has the same power at the ballot box as the smallest.

As WRCs are intergovernmental treaties, the private sector will not be able to vote on the outcome of the conference. However, the majority of regional preparatory meetings are open to the private sector, and the ITU-R study groups include paid-up 'associate' members of the ITU-R, so there are opportunities for the private sector to lobby for their desired outcome in the conference. Additionally, national regulators often consult on their conference positions beforehand.

Civil society, however, has less of an opportunity to participate in the decision-making process, even though the ITU is steadily allowing the general public greater access to its working documents.

Glossary

ABBREVIATION OR TECHNICAL TERM	MEANING
3GPP	Third-Generation Partnership Project: a body which brings together seven telecoms standard development organisations to produce common standards for cellular network technologies such as LTE, GSM, EDGE and HSPA
Third-Generation Partnership Project	See *3GPP*
Active sharing	Extensive sharing of facilities at a base station between two or more mobile operators. This will not only include the physical infrastructure described under *Passive Sharing* but also include elements of the electronics and RAN which make up the mobile network
Activity rules	Auction Rule which requires participants to be active bidders. These are designed to ensure the auction closes in a reasonable time frame and discourages strategic 'non-bidding'. In their simplest form, activity rules just require participants to make a bid in every round (see also *waiver*)
Administered incentive pricing	See *AIP*
Administrative assignment	Where a regulator chooses who should get spectrum licences by judging applications on the basis of a range of factors including the financial strength of the company, its technical expertise, its experience, its business plan and the amount it offers to pay. This is usually contrasted with assignment by auction, where the licence goes to the company which offers the highest amount of money

Continued

ABBREVIATION OR TECHNICAL TERM	MEANING
Aerial	A device to convert an electrical signal into an electromagnetic wave and vice versa
Aggregation	The desire to win separate lots in an auction in order to carry out your service (e.g. a mobile operator may need paired spectrum or several lots next to each other, contiguous spectrum, in order to have a large enough band to operate their service)
Aggregation risk	In multiple lot auctions where a bidder wins some but not all of the licences needed. This could happen where the auction is for several bands of spectrum and you need four of them to provide a service. A further issue is that you may need the bands to be next to each other (called 'contiguous spectrum')
AIP	Administered incentive pricing: Incentivising efficient usage by setting a spectrum price based on opportunity cost. Most frequently used for public sector spectrum (similar to opportunity cost pricing)
Allocation	The designation of a particular frequency band for a service type in the Radio Regulations
Analogue	Using continually varying rather than discrete values
Antenna	A device to convert an electrical signal into an electromagnetic wave and vice versa
ASA	Authorised Shared Access: see *LSA*
Ascending auction	Where prices go up in each round
ASMG	Arab Spectrum Management Group: group of national administrations in the ITU-R
Assignment	Giving an organisation a licence to use a specific frequency range
Assignment round	The final stage of a Combinatorial Clock Auction where participants can bid for specific frequency slots. In the first two stages participants bid for a generic licences in a particular band
ATU	African Telecommunications Union: specialized agency of the African Union, which is concerned with telecommunications
Band edge mask	The limit of emissions allowed within a particular band, which may include several blocks
Beauty contest	Popular term for administrative assignment (see *administrative assignment*)
BEREC	Body of European Regulators for Electronic Communication: Riga-based EU body that contributes to consistent application of EU law
Bidder behaviour rules	Restrictions on auction participants communicating with each other. These are used to prevent collusion

Continued

ABBREVIATION OR TECHNICAL TERM	MEANING
Bidder credits	These mean a bidder has a certain amount added onto their bid. For example, a participant with $10,000 of bidder credits making a winning bid of $50,000 would only need to pay $40,000 for the licence. Bidder credits are used in the United States to support companies promising to bring telecoms services to underserved areas, like Native Indian reservations
Block edge mask	The limit of emissions allowed within a particular block of frequencies (or a specific channel)
Carrier aggregation	The ability to operate a mobile network over two separate bands (e.g. 1800 MHz and 2.6 GHz). This is one of the features of LTE-Advanced
CCA	See *Combinatorial Clock Auction*
CEPT	European Conference of Postal and Telecommunications Administrations: Copenhagen-based group that facilitates pan-European co-operation on spectral and postal matters
CITEL	Inter-American Telecommunication Commission: entity of the Organization of American States that is concerned with telecommunications
Clock auction (Modern)	Participants indicate their demand at the stated price and the seller adjusts the price at regular intervals until demand equals supply
Clock auction (Traditional)	A clock auction is where the auctioneer cries out prices at regular intervals. This could be an ascending auction (prices go up each round) or the reverse, a descending auction
Clock Proxy Auction	Developed by three academics: Ausubel, Cramton and Milgrom. This is similar to a Combinatorial Clock Auction
Collusion	Where two or more bidders work together to keep the price down or damage a competitor, or consortia are formed prior to an auction limiting participation and hence lowering competitive pressures
Combinatorial auction	Auction where bidders can submit multiple bids on different combinations or 'packages' of licences
Combinatorial Clock Auction	The first two phases are a clock auction (modern) followed by a sealed bid stage. The optimal combination of all these generic bids which yields the highest value to the auctioneer is calculated by a computer program with the winners paying the second price. There is then an assignment round where participants can bid for specific frequency bands
Command and control	Where spectrum managers rather than the market decide upon the use of frequencies and who should have licences
Common value auction	The item in the auction has a common value to all bidders (e.g. an oil tract). However, bidders may hold different valuations due to different information in their possession

Continued

ABBREVIATION OR TECHNICAL TERM	MEANING
Compatibility studies	An assessment of the likelihood and extent of interference between different technologies or uses. These are a key part of the regulatory process in deciding spectrum allocations
Contiguous	Licences which are next to each other and provide a continuous band when put together
DAB	Digital audio broadcasting: a digital technology which aims to replace FM analogue radio
dB	deciBel: A representation of the ratio between two values, calculated using the logarithm (in base 10) of the ratio
De-modulator	A device to retrieve the information contained within an electrical signal
Decade	A factor of 10 when measuring frequencies
Descending auction	Where prices start high and go down in each round. See also *Dutch Auction*
Digital	Using discrete rather than continually varying values
DSA	Dynamic Spectrum Access: the use of advanced technology to dynamically identify and exploit unused frequencies
Duplex	Simultaneous two-way communication
Duplex spectrum	Another name for *Paired spectrum*
Dutch Auction	Seller sets a price and cries out lower and lower prices until a bidder accepts. The highest bidder wins and pays the bid amount. This is also known as open descending bids auctions
DVB	Digital Video Broadcasting: family of broadcasting standards developed by a European consortium
DVB-T	The terrestrial version of the DVB digital TV standard
EC	European Commission: the executive body of the European Union
ECC	Electronic Communications Committee: business committee of CEPT that establishes European consensus on technical matters relating to spectrum
EECMA	European Electronic Communications Market Authority: short-lived initiative to a create a single European spectrum regulator
EHF	Extremely High Frequencies (i.e. 30–300 GHz)
Electromagnetic spectrum	A complete set of frequencies incorporating radio as well as light, ultraviolet, infrared, x-rays and gamma rays
Eligibility points	Where each licence in an auction is assigned a number of points proportionate to the value of the licence. Bidders are given eligibility points based on the size of their deposit. This prevents bidders disguising their true intentions by bidding on low-value licences when they really want high-value ones
English auctions	The price is steadily raised by the auctioneer with bidders dropping out once the price becomes too high. This continues until there remains only one bidder who wins the auction at the current price. This is also known as open ascending-bid auctions

Continued

ABBREVIATION OR TECHNICAL TERM	MEANING
Exposure risk	The same as *Aggregation risk*
FCC	Federal Communications Commission: US communications regulator
Field strength	The strength of an electromagnetic wave; measured in Volts per metre
Filter	A device which separates one portion of the radio spectrum from another
First-price sealed bid	Bids are submitted confidentially: highest bidder wins and pays bid amount
Frequency	The unit used to distinguish different parts of the electromagnetic spectrum
FSS	Fixed Satellite Services: satellite services for stationary users on earth, such as TV
Galileo	European Union project to create a Global Navigation Satellite System that can compete with GPS and Glonass
GEO	Geostationary Orbit (referring to satellite services)
GHz	Giga Hertz – a billion Hertz
Glonass	Globalnaya Navigatsionnaya Sputnikovaya Sistema: a Global Navigation Satellite System that is managed by the Russian Federal Space Agency
GNSS	Global Navigation Satellite Systems: A global constellation of around 30 satellites at MEO that transmits signals that allow terrestrial users to determine their location
GPS	Global Positioning System: a Global Navigation Satellite System that is operated by the US Air Force.
GSM	Second-generation digital standard for cellular mobile phones. The initials originally stood for Groupe Spécial Mobile but were also translated as Global System for Mobile Communications
GSMA	GSM Association: An interest group that represents the mobile industry
Guard band	A piece of spectrum purposefully left between two radio signals to prevent them from interfering with each other
HD	High Definition: a TV picture which has upwards of about 0.9 to 2.07 megapixels
HF	High frequencies [i.e. 3000–30,000 kHz (3–30 MHz)]
Hz	Hertz: the unit of frequency
ICT	Information and Communications Technologies: a term widely used to reflect the interdependence of the telecoms and computing markets
In-band emissions	Level of emissions allowed within the range of frequencies assigned to a user
Interference	Interruptions or degradations to one service caused by radio signals from a different service

Continued

ABBREVIATION OR TECHNICAL TERM	MEANING
Interleaved spectrum	Where a frequency used by a transmitter in one part of the country is effectively unoccupied outside, the range of that transmitter in another part of the country
IoT	Internet of Things: Internet connections between machines, both large and very small, often using wireless technologies (see also *M2M*)
IP	Internet Protocol
IPTV	Internet Protocol Television (i.e. TV delivered over the Internet)
ITU	International Telecommunications Union: Geneva-based specialised agency of the United Nations for Information and Communication Technologies
ITU-D	Section of the ITU dedicated to improving ICTs in developing countries
ITU-R	The Radio Bureau of the ITU
ITU-T	Section of the ITU dealing with international standards
Jamming	Deliberately creating interference
kHz	kilo Hertz – a thousand Hertz
LF	Low frequencies (i.e. 30–300 kHz)
Licensed Shared Access	See *LSA*
LSA	Licensed Shared Access: a spectrum sharing agreement with specified parameters which can be defined and enforced by software. Very similar to Authorised Shared Access (ASA)
LTE	Long Term Evolution; often referred to as 4G. This follows on from GSM and UMTS technologies to offer higher data rates and other improvements
LTE-Advanced	An enhanced version of LTE offering features such as carrier aggregation. It was standardised in March 2011 as 3GPP Release 10
M2M	Machine-to-machine communication – similar to the Internet of Things (IoT)
Man-made noise	Noise on radio frequencies generated by electrical equipment
Massive MIMO	Where base stations have a large number of antennas able to form a large number of beams (See also *MIMO*)
MF	Medium frequencies (i.e. 300–3000 kHz)
MHz	Mega Hertz – a million Hertz
MIFR	Master International Frequency Register: an ITU-R document that lists all authorised uses of frequencies for broadcasting and for satellites
Millimetre waves	See *mmWaves*
MIMO	Multiple Input–Multiple Output: using multiple antennas to form directional beams. The reflections caused by objects between the transmitter and receiver allow the reuse of the same frequency multiple times

Continued

ABBREVIATION OR TECHNICAL TERM	MEANING
mmWave	Millimetre wave: strictly speaking frequencies between 30 and 300 GHz, but commonly used to refer to frequencies far higher than those currently used for IMT
Modulation	The act of encoding information onto a radio signal
MSS	Mobile Satellite Services: satellite services for mobile communication services on earth
MVNO	Mobile Virtual Network Operator: a wholesale agreement to use mobile networks owned by other companies
Natural noise	Noise on radio frequencies caused by natural occurrences (such as lightning storms)
Noise	Radio signals which appear on a frequency without any transmissions being present
Noise floor	The lowest level of noise found on any particular frequency. Usually a combination of man-made and natural noise
NTIA	National Telecommunications & Information Administration: US executive branch agency that advises government on communication issues and regulates the public sector use of spectrum
OFDM	Orthogonal frequency division multiplexing
Opportunity cost	The value forgone when a resource such as a frequency is used for one purpose rather than another
Opportunity cost pricing	Pricing the spectrum at (or close to) the next highest value that another user would have paid (e.g. if the government is using a band which could be used by mobile industry, the charge to the government should equal what a mobile operator would have been prepared to pay). Similar to *Administered Incentive Pricing*
Out-of-band emissions	Level of emissions allowed outside the range of frequencies assigned to a user
Package bidding	This refers to auctions where bidders may place bids on groups of licenses as well as on individual licenses. This allows bidders to place a higher value on winning a group of licences (perhaps because they are contiguous or in complementary pairs). Package bidding allows participants to bid on a whole group of licences or none at all, thereby avoiding withdrawal penalties if they cannot win some of the desired licences. Package bidding can help prevent *Aggregation risk*
Paired spectrum	Two pieces of spectrum, one used for communications in one direction and the other in the opposite direction (e.g. to and from a mobile handset)
Passive sharing	Mobile operators have joint use of non-radio facilities at a base station, such as the site compound, masts and cabinets, air-conditioning and power supply. Usually contrasted with *Active sharing* (see separate entry)
PCAST	US President's Council of Advisors on Science and Technology

Continued

ABBREVIATION OR TECHNICAL TERM	MEANING
PMR	Private mobile radio: a mobile system which is only for the use of the licensee or their company (e.g. business radio used by a taxi company). Usually contrasted with *Public mobile radio* (see separate entry)
Price discovery	Where an auction bidder can see how much the other participants are willing to pay for the spectrum (e.g. in a multiple round auction where the prices are published after each round). Price discovery is supposed to reduce the risk of overvaluing the spectrum
Private mobile radio	See *PMR*
Proxy auction	Where someone submits a bid on your behalf. The proxy could also be a computer rather than a human (e.g. eBay will submit bids on your behalf up to a specified amount)
Public Mobile Radio	A mobile system which any member of the public can use, unlike private mobile radio which is restricted to the organisation which holds the licence. Public Mobile Radio is the ITU service type which includes mobile cellular services
QoS	Quality of Service: the standard required by a spectrum user – often defined as part of sharing or wholesale agreements
Quality of Service	See *QoS*
Radio astronomy	The study of celestial objects such as stars and galaxies by monitoring the radio waves which they emit. Radio astronomers are very unusual among spectrum users because they are only concerned with receiving signals but not transmitting them
Radio Regulations	Rules on the use of radio spectrum that all national administrations in the ITU have agreed to
Radio spectrum	The portion of the electromagnetic spectrum that yields radio waves. Typically viewed as 0–1000 GHz
RAN	Radio Access Network: the wireless portion of a mobile network
RCC	Regional Commonwealth in the field of Communications: Moscow-based body that enables co-operation among members of the Commonwealth of Independent States (CIS)
Re-farming	Replacing one use or user of the radio spectrum with a different one, normally over a complete frequency band
Receiver	A device which converts an incoming radio signal into information
ROES	Receive Only Earth Station (a satellite facility)
RSPP	Radio Spectrum Policy Programme: a series of initiatives on spectrum in the European Union
SAB	Services Ancillary to Broadcasting – devices used to support broadcast activities such as radio microphones and wireless cameras
SD	Standard Definition: a TV picture which has about 0.3 megapixels
Sealed bid	When bids are submitted in sealed envelopes so nobody knows what their opponents have bid (see *Price discovery*)

Continued

ABBREVIATION OR TECHNICAL TERM	MEANING
Second-price sealed bid	Bids all submitted confidentially: highest bidder wins and pays second highest bid (also known as *Vickery Auction*). Designed to prevent the winner's curse
Shannon limit	The maximum amount of data that can be carried by any communication channel
Shannon–Hartley theorem	A theory that connects data capacity, signal-to-noise and bandwidth and yields the Shannon limit
SHF	Super high-frequencies [i.e. 3000–30,000 MHz (3–30 GHz)]
Signal-to-noise ratio	The ratio (usually measured in decibels) between a wanted signal and any noise or interference present on the same radio frequency
Simultaneous Multiple Round Auction	In this auction, all licenses are available for bidding throughout the entire auction. SMRAs have discrete, successive rounds, with the bids made public after each round. SMRA lots are licences for specific frequency bands, rather than generic lots as in the first two stages of a Combinatorial Clock Auction
SMRA	See *Simultaneous Multiple Round Auction*
Software radios	Radio receivers (and transmitters) which use software to handle the modulation and de-modulation
Spectrum	Normally used to mean the radio spectrum, though could also represent the whole electromagnetic spectrum
Spectrum cap	A limit on the amount of spectrum a bidder can buy in an auction. Usually implemented through eligibility points (see above)
Spectrum mask	A graphical representation of the in-band and out-of-band emission limits
Spectrum usage right	Spectrum usage right: a type of licencing used in the United Kingdom based on measuring interference to receivers rather than setting limits on transmissions
Spread spectrum	A technology which spreads a radio signal over more spectrum than is necessary
Strategic behaviour	Where participants adopt bidding behaviour that is intended to send misleading signals to rivals about valuations and demand
Substitution risk	Paying a high price for some lots when substitute lots could have been purchased at a lower price. The ability to switch lots in SMRAs can prevent this
SUR	See *Spectrum usage right*
Switching rules	Allow bidders to move to substitutable lots as the price rises. Encourage aggregation
Technology-neutral licences	Licences which allow any technology to be used in a frequency band, provided it meets specified technical criteria. These have largely replaced earlier licences which controlled interference by specifying a technology and making calculations based on its technical characteristics

Continued

ABBREVIATION OR TECHNICAL TERM	MEANING
THF	Tremendously High Frequencies (i.e. 300–3000 GHz)
Transaction cost	A cost incurred in making an economic exchange or participating in a market
Transmission	The emission of electromagnetic waves
Transmitter	A device which generates electrical signals at radio frequencies
UHDTV	Ultra-high-definition television: usually divided into 4K UHDTV where the picture has 8.29 megapixels or 8K UHDTV with 33.18 megapixels
UHF	Ultra High Frequencies (i.e. 300–3000 MHz); however, people often use UHF to refer to what is in fact a portion of the band, namely, 470 to 862 MHz. This was allocated to broadcasting in Region 1 (and in many other parts of the world) and is more accurately known as UHF Bands IV and V
UHF broadcast band	This is often used to refer to 470 to 862 MHz, all or some of which is used for terrestrial broadcasting. The technical term for this band is UHF Bands IV and V
UMTS	Universal Mobile Telecommunications System: a third-generation mobile cellular system following on from the GSM standard
VHF	Very High Frequencies (i.e. 30–300 MHz)
Vickery Auction	See *Second-price sealed bid*
VLF	Very Low Frequencies [i.e. 3000–30,000 Hz (3–30 kHz)]
Waiver	In an auction, a waiver, if allowed, is where a bidder places no bid but remains active in the auction. Waivers are supposed to allow bidders time for reflection and possible communication with financiers. Auction participants may typically submit two or three waivers, though increasingly their use is not seen as helpful
WAPECS	Wireless Access Platform for Electronic Communication Services Policy: an attempt by the European Commission to create a spectrum mask which would allow the deployment of current and foreseeable future technologies in the main mobile bands
Waveform	An electrical signal representing a radio wave
Wavelength	The length of an electromagnetic wave, usually measured in metres
Wi-Fi	Wireless Local Area Network (WLAN) technology that is based on IEEE 802.11 standards
WiGig	Wi-Fi-like devices which operate in the 60 GHz band but offer higher data rates and can be used to replace cables, for wireless docking between devices like laptops and tablets and for streaming HD videos
Winner's curse	Where the winner of the auction realises that they have bid more than the value of the item
WRC	World Radio Conference: intergovernmental event held by the ITU every 3 or 4 years in which the Radio Regulations can be amended

Index